# Energy Harvesting

***Energy Harvesting: Enabling IoT Transformations*** gives insight into the emergence of energy-harvesting technology and its integration with IoT-based applications. The book educates the reader on how energy is harvested from different sources, increasing the effectiveness, efficiency and lifetime of IoT devices.

- Discusses the technology and practices involved in energy harvesting for biomedical, agriculture and automobile industries
- Compares the performance of IoT-based devices with and without energy harvesting for different applications
- Studies the challenges and issues in the implementation of EH-IoT
- Includes case studies on energy-harvesting approaches for solar, thermal and RF sources
- Analyzes the market and business opportunities for entrepreneurs in the field of EH-IoT.

This book is primarily aimed at graduates and research scholars in wireless sensor networks. Scientists and R&D workers in industry will also find this book useful.

## Chapman & Hall/CRC Internet of Things: Data-Centric Intelligent Computing, Informatics, and Communication

The role of adaptation, machine learning, computational intelligence, and data analytics in the field of IoT systems is becoming increasingly essential and intertwined. The capability of an intelligent system is growing depending upon various self-decision-making algorithms in IoT devices. IoT-based smart systems generate a large amount of data that cannot be processed by traditional data processing algorithms and applications. Hence, this book series involves different computational methods incorporated within the system with the help of analytics reasoning, learning methods, artificial intelligence, and sense-making in big data, which is most concerned with IoT-enabled environments.

This series aims to attract researchers and practitioners who are working in information technology and computer science in the fields of intelligent computing paradigms, big data, machine learning, sensor data, the Internet of Things, and data sciences. The main aim of the series is to make available a range of books on all aspects of learning, analytics, and advanced intelligent systems and related technologies. This series will cover the theory, research, development, and applications of learning, computational analytics, data processing, and machine learning algorithms as embedded in the fields of engineering, computer science, and information technology.

### Series Editors:
### Souvik Pal
Sister Nivedita University, (Techno India Group), Kolkata, India
### Dac-Nhuong Le
Haiphong University, Vietnam

**Security of Internet of Things Nodes: Challenges, Attacks, and Countermeasures**
*Chinmay Chakraborty, Sree Ranjani Rajendran and Muhammad Habib Ur Rehman*

**Cancer Prediction for Industrial IoT 4.0: A Machine Learning Perspective**
*Meenu Gupta, Rachna Jain, Arun Solanki and Fadi Al-Turjman*

**Cloud IoT Systems for Smart Agricultural Engineering**
*Saravanan Krishnan, J Bruce Ralphin Rose, NR Rajalakshmi, N Narayanan Prasanth*

**Data Science for Effective Healthcare Systems**
*Hari Singh, Ravindara Bhatt, Prateek Thakral and Dinesh Chander Verma*

**Internet of Things and Data Mining for Modern Engineering and Healthcare Applications**
*Ankan Bhattacharya, Bappadittya Roy, Samarendra Nath Sur, Saurav Mallik and Subhasis Dasgupta*

**Energy Harvesting: Enabling IoT Transformations**
*Deepti Agarwal, Kimmi Verma and Shabana Urooj*

# Energy Harvesting
## Enabling IoT Transformations

Edited by
Deepti Agarwal
Kimmi Verma
Shabana Urooj

**CRC Press**
Taylor & Francis Group
Boca Raton  London

CRC Press is an imprint of the
Taylor & Francis Group, an **informa** business

A CHAPMAN & HALL BOOK

First edition published 2023
by CRC Press
6000 Broken Sound Parkway NW, Suite 300, Boca Raton, FL 33487–2742

and by CRC Press
4 Park Square, Milton Park, Abingdon, Oxon, OX14 4RN

*CRC Press is an imprint of Taylor & Francis Group, LLC*

ISBN: 978-1-032-11183-4 (hbk)
ISBN: 978-1-032-34925-1 (pbk)
ISBN: 978-1-003-21876-0 (ebk)

DOI: 10.1201/9781003218760

Typeset in Minion
by Apex CoVantage, LLC

# Contents

# Preface

With the emergence of Internet of Things (IoT) technology, the effectiveness and efficiency of wireless sensor nodes (WSNs) is becoming one of the most critical factors to be considered by applications such as traffic management, healthcare systems, environmental monitoring and smart buildings. The future of upcoming IoT deployments relies on the capabilities of nodes to operate without any interruption utilizing ambient energy harvesting (EH).

Over the years, the main hurdle in the growth of IoT-enabled devices has been identified as power for interconnected nodes. When connected nodes interact with each other, they consume a large amount of energy for effective communication, and because of this, the devices operate only for a limited duration. A promising alternative with no dependency on battery life is energy harvesting. EH is the process of taking energy from various different sources, such as solar energy, wind energy, thermal energy, radiant energy (radio frequency), mechanical energy (motion), finer motion, footfalls, breathing, vibrations and so on. Once energy is received from these sources, it is accumulated and further converted into usable electrical energy. This harvested electrical energy is used to power IoT devices and also increases their overall efficiency, durability and lifetime. Though energy harvesting plays a very significant role by increasing the efficiency and lifetime of IoT devices, EH systems also have some limitations, such as a low amount of energy harvested, dependency on the source from which the energy is to be harvested, inefficiency of the harvesting system and so on. In addition, there is one critical problem: power management, which impacts battery storage. In order to mitigate this problem, power-management integrated circuits were introduced to provide better power management to IoT devices. In the last decades, efforts have been made to overcome these limitations, and new models of energy harvesting have been introduced.

Governments and public-sector organizations have a key role to play in promoting energy-harvesting techniques and their significance when used in combination with IoT devices. In order to boost the prospects, all such companies developing energy harvesters need the funding, support and investment in research and development (R&D) that governments should provide. There are still many product-based organizations working on IoT devices that do not recognize the principle of energy harvesting or the benefits they could achieve by using this technology. Organizations should allot a sufficient marketing budget and drive public awareness programs to educate and emphasize the significance of EH systems and the benefits they provide when combined with IoT devices.

This book demonstrates the emergence of energy-harvesting technology and its integration with IoT-based applications. It also focuses on efficient utilization of energy harvesting in IoT-based devices in the fields of agriculture, health devices, industry automation and smart traffic management in order to minimize human effort. This will improve industrial systems for efficiency in commercial and industrial areas as well. The book will discuss the technical requirements as well as the implementation of IoT-based devices used with low power consumption and will cover the application of EH-IoT in various industries.

The main focus of this book is not limited to wireless communication. It also aims to demonstrate the benefits of implementing various models of the IoT with energy-harvesting approaches for solar, mechanical, thermal and vibrational sources. The book also presents energy harvesting–based IoT applications for entrepreneurial opportunities.

This book showcases the future of self-powered IoT devices, with contributors from around the world. The contributing authors have expertise extending from research scholars, scientists and R&D employees from industry to academicians and postgraduate students. The concepts and applications discussed in this book will be the foundation of technologies that will be realized in the next few years.

# Editor Biographies

**Dr. Deepti Agarwal** received her PhD from Indira Gandhi Delhi Technical University for Women (IGDTUW), Delhi, in 2019. She received an MTech in energy and environmental management from the Indian Institute of Technology Delhi (IITD), New Delhi, in 2009 and a BTech in electronics and communications engineering from Uttar Pradesh Technical University (UPTU), Lucknow, in 2005. She worked as an assistant professor in Mahatma Gandhi Mission's College of Engineering & Technology, Noida, India, from 2005 to 2014. She is currently working as an associate professor on the Delhi Technical Campus, Greater Noida (affiliated with GGSIPU). She is an active member of international societies like IEEE, ISTE and InSc. She has successfully completed many reviews for highly reputed journals like *IEEE/OSA Applied Optics*, *IEEE/OSA JOCN* and *IET Communications* in the field of engineering. She has published more than 15 research papers in refereed journals and IEEE conferences with SCI and Scopus indexing. Her research interests include free-space optics, cooperative communications, unification of channel models and differential non-coherent networks.

**Dr. Kimmi Verma** has been an academician for the last 17 years in the field of electronics engineering. She obtained her PhD in biomedical image processing from the Department of Electronics Engineering (Banasthali University) and did her MTech in microelectronics (UPTU) and BTech in electronics and instrumentation engineering (UPTU). She is a Class A Certified

Professional in Robotics under the MHRD project E-yantra from IIT Mumbai. She has teaching experience of more than 16 years in engineering academics. She has authored and co-authored several research papers published in high-quality international journals and reputed conference proceedings. She has successfully completed editorial responsibilities for reputed journals. She has three patents in the field of electronics and advanced computing technologies. She is the author of *Analog and Digital Electronics* with University Science Press, and she is currently working as an associate professor at IP University, a reputed college.

 **Dr. Shabana Urooj** (Senior Member, IEEE) received her BE in electrical engineering and MTech in electrical (instrumentation and control) from Aligarh Muslim University, Aligarh, UP, India, in 1998 and 2003, respectively. She obtained her PhD in biomedical instrumentation from the Electrical Engineering Department, Jamia Millia Islamia (A Central University), Delhi, India. She has served in industry for three years and teaching organizations for more than 20 years. Presently she is working as Associate Professor at the Department of Electrical Engineering, College of Engineering, Princess Nourah bint Abdulrahman University, Saudi Arabia, on leave from Gautam Buddha University, UP, India. She has guided seven doctorates and more than 65 MTech dissertations. She has authored and co-authored more than 200 research articles, which are published in high-quality international journals and reputed conference proceedings. She has successfully completed editorial responsibilities for reputed journals and several quality books and proceedings. She has served as Associate Editor of reputed journals, such as *IEEE Sensors Journal, Frontiers Energy Research (Smart Grid)* and special issues of *MDPI*. She was a recipient of the Springer's Excellence in Teaching and Research Award, the American Ceramic Society's Young Professional Award, the IEEE's Region 10 Award for Outstanding Contribution in Educational Activities, Research Excellence Award for quality publishing from Princess Nourah University and several best paper presentation awards. Dr. Urooj is serving as an active volunteer at the Institute of Electrical and Electronics Engineering in various capacities. She has received the title of IEEE STEM Ambassador. She has been associated with the IEEE Delhi Section in various volunteer positions for more than ten years.

# Contributors

**Piyush Agarwal**
Dehradun, Uttarakhand,
India

**Shafeeq Ahmad**
Al Falah University, Haryana
India

**Md. Toufique Alam**
Al Falah University, Haryana
India

**A. Ambikapathy**
Galgotias College of Engineering
and Technology, Greater Noida
India

**Ashish**
Integral University Lucknow,
Uttar Pradesh
India

**Vikas Singh Bhadoria**
ABES Engineering College,
Ghaziabad, Uttar Pradesh,
India

**Mohammad Bilal**
College of Eng. and Technology,
Greater Noida, Uttar Pradesh
India

**Ankit Gambhir**
Delhi Technical Campus,
Greater Noida, Uttar Pradesh
India

**Rathishchandra R. Gatti**
Sahyadri College of Eng. &
Management,
Mangaluru, Karnataka
India

**Tarikul Islam**
Jamia Millia Islamia,
New Delhi
India

**Shiva Pujan Jaiswal**
SET, Sharda University
Gr. Noida,
Uttar Pradesh, India

**Mohd Zaheen Khan**
IET Lucknow, Uttar Pradesh
India

**Osama Khan**
Jamia Millia Islamia, Delhi
India

**Sandeep Kumar**
Central Research Laboratory, BEL,
Ghaziabad, Uttar Pradesh, India

**Sonam Lata**
Jamia Millia Islamia,
Delhi, India

**Priya Matta**
India Department of CSE, Graphic
    Era,
Dehradun, Uttarakhand,
India

**Shabana Mehfuz**
Jamia Millia Islamia, Delhi
India

**Mahak Narang**
Delhi Technical Campus,
Greater Noida, Uttar Pradesh
India

**Arshi Salamat**
Jamia Millia Islamia,
New Delhi
India

**Sachin Sharma**
Dehradun, Uttarakhand,
India

**Vivek Shrivastava**
NIT, New Delhi,
India

**Mandeep Singh**
Thapar Institute of Engineering
    and Technology,
Patiala, Punjab
India

**Punit Kumar Singh**
Integral University
Lucknow, Uttar Pradesh
India

**Ranjeeta Singh**
Engineering, SET, Sharda
    University,
Gr. Noida, Uttar Pradesh, India

**Sudhakar Singh**
Lovely Professional University,
Phagwara, Punjab
India

**Hassan Usman**
Integral University Lucknow,
Uttar Pradesh
India

**Poonam Yadav**
NERIST, Nirjuli,
Arunachal Pradesh, India

# CHAPTER 1

# Energy Harvesting–Based Architecture in IoT

## Basics of Energy Harvesting, Key Technology for Enhancing the Life of IoT Devices, Challenges of IoT in Terms of Energy and Power Consumption

Sandeep Kumar, Poonam Yadav

## CONTENTS

## 1.1 INTRODUCTION

The Internet of Things (IoT) is a new computer paradigm that aims to turn commonplace objects into smart ones. It has emerged as one of the preeminent technologies of the current century, with applications in transportation, civil infrastructure, geological monitoring, healthcare, defence, manufacturing, and industry [1]. The premier idea for IoT was proposed by Massachusetts Institute of Technology researchers in 1999

DOI: 10.1201/9781003218760-1

[2]. However, in 2005, at the World Summit on Information Society held in Tunisia, this concept was formally accepted by the International Telecommunication Union (ITU) by releasing two reports [3]. These reports defined the IoT as an emerging paradigm that can support multiple connections and create a network of networks by connecting the physical and digital worlds. According to expert predictions, by 2050, about 22 billion IoT devices might be connected to the internet and communicating through the IoT environment [4]. Currently, the IoT is in charge of maintaining capillary networking infrastructure, which includes a significant number of devices connected to the internet and operating from anywhere at any time. IoT devices are nonstandard computing devices that are powered by batteries and may connect to a network wirelessly to send and receive data. The finite capacity of the batteries becomes a severe barrier when it comes to the efficiency of IoT devices because when IoT devices connect with each other, a large amount of energy is used, resulting in devices operating for only a short period of time, as long as the battery lasts. For small IoT systems, replacement of batteries can be a solution, but it is ineffective for large systems. Energy harvesting (EH) is a viable answer to this problem. EH is a technique to power up these devices, acquiring energy from the surrounding operational environment. This in turn provides a promising future for self-sustainable wireless networks by eliminating the dependency on battery charging and deployment in hard-to-reach places.

This chapter's goal is to emphasise the potential of energy-harvesting technologies to minimise reliance on fixed charge batteries and improve device longevity, resulting in uninterrupted device use. The chapter discusses the fundamental components of EH systems, figures out prospective energy resources, and estimates the potential of different harvesting techniques.

## 1.2 THE IoT NETWORK AND ITS COMPONENTS

The IoT entails a complex ecosystem of tools and services that must work together to provide a comprehensive solution. The IoT is intended to take connectivity to the next level by connecting several devices to the internet at the same time, allowing for easier man–machine and machine–machine interactions. The major components of the IoT ecosystem are

(i) *IoT devices*: These comprise transducers/sensors and actuators, which are responsible for collecting data from the surroundings.

(ii) *IoT connectivity*: The collected data is transferred in a cloud infra-structure using different communication media such as such as Wi-Fi, Bluetooth, cellular networks, and many more.

(iii) *IoT cloud server*: The data sent over the gateway is safely saved and processed on the cloud server, which is located in data centres. This processed data is then used to take intelligent action, making all of our devices smart. All analytics and choices are made in the cloud.

(iv) *End-user devices and user interfaces*: The data is made available to the user via a user interface such as text messages, email, alerts, and so on.

During the communication of IoT devices, a huge amount of energy is consumed, which limits their efficiency and lifetimes. EH has emerged as a key solution to this problem. The first step to enabling EH is to create an ultra-low-power connectivity solution. If a device can operate on harvested energy alone, the need for charging or a permanent power cord is eliminated. By combining wireless technology and energy-harvesting solutions, true wireless—as in no power cord needed—is a possibility for IoT devices.

## 1.3 REQUIREMENTS OF EH IN IoT ECOSYSTEMS

As mentioned, IoT networks are battery-operated networks, and batteries have finite capacity and lifetimes. To realise the dream of smart living, deployment of IoT devices is accelerating in every field. All the components of the IoT ecosystem are power hungry and are continuously running to transmit and receive data. While exploring the application of IoT devices, it is required to recognise the power density requirements of the IoT device, as devices equipped in the IoT infrastructure have their own unique power requirements. Therefore, the energy harvester must be able to scavenge at least the minimum power that is required by the devices in the IoT ecosystem. Even though breakthroughs have been made in optimising power management systems, communication protocols, and operating systems, energy will eventually get depleted. It often becomes difficult to replace and maintain when devices are placed in inaccessible locations [5]. Including this, there are several other limitations of battery-operated IoT devices, such as

(i) The application area of batteries is limited, as functioning in harsh environments leads to capacity and power losses [6].

(ii) The weight and dimensions of the battery have a direct effect on its capacity.

(iii) Another important worry is the environment, as batteries contain hazardous chemicals and poisons, making disposal more difficult.

Although many initiatives have been taken to speed up the recycling rate of discarded batteries, improvements in this direction still require critical attention to achieve sustainable development goals. These reasons define the need for EH implementation in the design of current wireless networks supporting IoT applications.

From Table 1.1, it can be seen that IoT devices have certain power ranges required for their operation [7–10]. As the key feature of an EH system is scavenging power from the ambient environment, Figure 1.1 shows the intermittent power available from some of the well-known ambient sources [51]. Power ranging from 0.1 μW to 100 W can be harvested from various energy sources. A thermo-electric generator (TEG) has the capability to harness energy ranging from 10μ W–100 W. This works on the principle of the thermoelectric effect, the temperature difference between two materials. Though its efficiency is very low, up to 2–5%, it can still power IoT devices even with a low thermal gradient. Harvesting solar energy using photovoltaic (PV) cells is acquiring a major place, as it has a wide power range, extending from microwatts to megawatts. Pyroelectric generators, which are again thermal-based sources, are still in the research phase, but they are expected to harvest power in the microwatt (μW) to milliwatt (mW) range. The energy harvested from RF sources provides limited production ranging from 0.1 μW–1 mW, which strongly depends upon the environment and the size of the rectenna, so this can be implemented in applications where size is not a major issue and life expectancy is beyond 15 to 20 years. Also, triboelectric energy is not commercially

TABLE 1.1    Power Range Requirements of Various IoT Devices and Sensors

| S. No | Electronic Module | Power Range |
|---|---|---|
| 1 | Radio Frequency Identification Tag | 10 μW |
| 2 | Sensor/Remote | 100 μW |
| 3 | Wireless Sensor/Hearing Aid | 1 mW |
| 4 | Bluetooth Transceiver | 10 mW |
| 5 | Global Positioning System (GPS) | 100 mW |

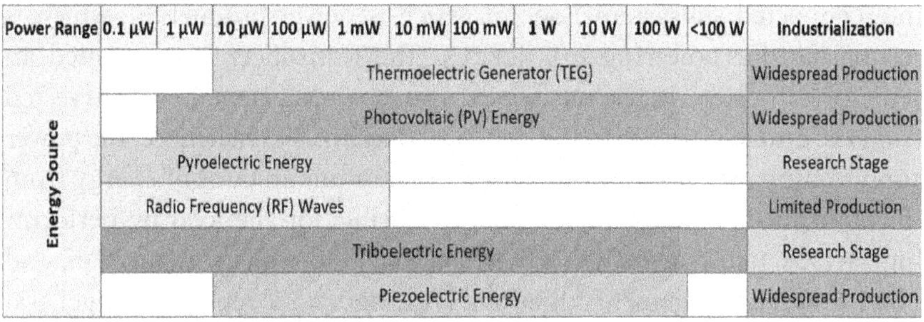

| Power Range | 0.1 µW | 1 µW | 10 µW | 100 µW | 1 mW | 10 mW | 100 mW | 1 W | 10 W | 100 W | <100 W | Industrialization |
|---|---|---|---|---|---|---|---|---|---|---|---|---|
| Thermoelectric Generator (TEG) | | | | | | | | | | | | Widespread Production |
| Photovoltaic (PV) Energy | | | | | | | | | | | | Widespread Production |
| Pyroelectric Energy | | | | | | | | | | | | Research Stage |
| Radio Frequency (RF) Waves | | | | | | | | | | | | Limited Production |
| Triboelectric Energy | | | | | | | | | | | | Research Stage |
| Piezoelectric Energy | | | | | | | | | | | | Widespread Production |

FIGURE 1.1 Power generated intermittently from common ambient energy sources

available and is still under exploration. Triboelectric energy is created by contact electrification due to frictional contact between different materials. This energy source is expected to produce energy between 0.1 µW and megawatt levels. Piezoelectric energy, further, produces a large amount of power, ranging from 10 to 100 watts. This type of energy has the potential to be massively industrialised. IoT systems equipped with EH techniques have numerous benefits over the conventional battery-driven system. Such as EH solutions can

- Lower the reliance on battery power, as well as installation and maintenance expenditures.

- Provide operating capabilities in inaccessible and hazardous environments.

- Reduce the negative impact on the environment. EH has the potential to eliminate the cost of batteries and the energy expenses associated with battery replacement.

## 1.4 ENERGY-HARVESTING TECHNOLOGY

EH basically converts ambient energy into electrical energy to power IoT devices. Our surroundings are full of unused ambient energy sources such as solar, wind, vibrations, RF waves, and so on, which are non-exhaustible and sustainable for almost an infinite time. Several mechanisms have been developed over the years to utilise ambient energy. With progress in technologies and materials, a wide spectrum of energy sources can be utilised to power IoT devices. An EH system is made up of numerous interconnected subsystems, one of which is power production, which is responsible for powering IoT devices. The transducer often termed an "energy harvester" is responsible for transforming the energy harvested from the ambient source to the electrical domain. Furthermore, the power conditioning unit serves to scavenge a maximum amount of energy from the transducer. Also, it makes energy feasible for the load by performing various adjustments like voltage rectification, voltage regulation, and other power management functions [11]. Figure 1.2 represents a general block diagram for an EH system. Basically, the EH process involves energy conversion hardware that converts the surrounding energy into electrical energy, which is then conditioned by a power management circuit, stored in energy storage elements, and eventually supplied to the electrical load.

Energy harvester architecture is classified in two classes: harvest-use (just-in-time) and harvest-store-use.

(i) Harvest-use architecture: In harvest-use architecture, the harvested energy is directly utilised to power up IoT devices. It avoids the usage of voltage converters and, in the long run, removes energy storage; as a result, the shortcomings of the harvest-store technique are successfully eliminated [12], lowering device costs and potentially enhancing system efficiency.

(ii) Harvest-store-use architecture: As the name explains, energy is harvested and stored for future use [12]. This system requires storage elements like super-capacitors and rechargeable batteries to store energy, and these need to be selected depending on the surrounding conditions.

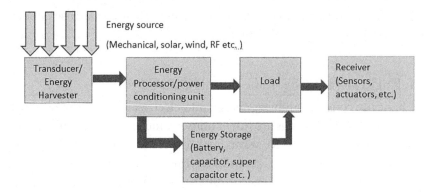

FIGURE 1.2   Block diagram of EH system

**Energy-Harvesting Resources** Many research and development projects are being carried out on harvesting energy from various renewable energy sources available in the environment. On the basis of the sources from which the energy can be harvested, EH systems are further classified into categories, as shown in Figure 1.3.

(A) *Ambient EH*: Ambient energy sources are abundantly available in the environment and can be harvested at no cost. These energy resources are further classified as solar, wind, radio frequency, and thermal-based energy resources.

    (i) *Solar EH*: Solar energy is the most abundant power available on the earth. Solar power can provide an indefinite amount of energy to power IoT devices. The total amount of solar energy reaching the earth surface is about 7,000 to 8,000 times more than the annual primary energy consumption across the globe [13–14]. Sunlight is harvested as shown in Figure 1.4, using solar or PV cells that are made of n-type and p-type semiconductor material. An electric field is developed at the junction of these, termed the p-n junction, and as it is exposed to light, electrons are released. The amount of solar energy converted into useable electric energy depends upon the efficiency of the PV cells used. During the transfer of the reaped solar energy from the harvester, a large amount of energy is lost. The maximum power point tracker (MPPT) circuit [15] has been proposed as a solution to this problem which can be implemented along with a power conditioning unit to proficiently transfer the harvested solar energy from the harvester to a rechargeable battery. For places where guaranteed sunlight is available, PV cells are the best harvesters. However, improvements in manufacturing techniques and circuits, as well as better IoT device designs, have permitted the adoption of indoor solar-powered device systems. The commonly available solar energy harvesters that are most widely used are Hydrowatch, Fleck, Solar Biscuit, Everlast, Enviromote, and others. Fleck is designed to operate in full sunlight conditions, using two super-capacitors [16]. Everlast [17] has a life span of 20 years, and it has a connection of a super-capacitor to a solar harvester and is independent of batteries. The Solar Biscuit [18] harvester is close to Everlast, but it is used in integrated mode with a super-capacitor and is directly connected to the cell without the need for a voltage regulator. The Hydrowatch [19] harvester uses a MPPT system, is a single-source harvester, and does not need software control for the battery, which is its best feature. The generated energy has a high efficiency, but due to the need for batteries for storage, the lifespan is limited.

(ii) *RF EH*: This techniques harvests the electromagnetic energy captured from different RF energy sources such as wireless internet, cellular network, radio and satellite stations, and so on. This form of energy exists at various levels all around us (indoors and out), and it is

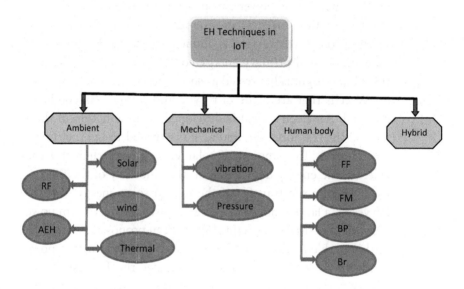

FIGURE 1.3   Taxonomy of EH techniques used in IoT ecosystems

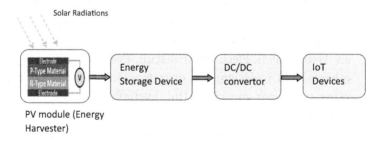

FIGURE 1.4   Schematic of solar energy-harvesting technology

always available. This energy can be collected indefinitely and promises tremendous scope to replace the small batteries in IoT systems. The key component of a RF EH system consists of an antenna and rectifier circuit that assists in converting RF power into DC power. Along with this, the EH system could include a power management block for energy storage. The energy storage subsystem is in charge of storing all captured energy as well as maintaining a consistent output voltage. Although it has low efficiency and power density [20], it emerges as a potential solution to power IoT devices in many scenarios [21–22]. In spite of the huge potential of ambient RF power scavenging, electromagnetic wave energy is extremely low and weakens as the signal moves away from the source. So, to gain maximum efficiency, the harvester must be close enough to the RF source. One feasible solution to this can be the use of a dedicated RF transmitter generating powerful RF signals just to power the sensors.

(iii) *Wind EH*: Wind energy is one of the best sustainable energy sources after solar energy. The energy of the air flow is harvested. To extract the energy from constant vibrations of the wind flow, piezoelectric devices can be used [23]. To implement an EH system with IoT components, design of miniature wind EH systems is still ongoing research. In [24], a wind energy harvester is used to power the sensors in a healthcare monitoring system.

(iv) *Acoustic Energy Harvesting*: Due to the plenitude of clean sound waves, a new technical innovation known as acoustic energy harvesting (AEH) or sound energy harvesting is being introduced. Acoustic energy can be transmitted via vibrational or sound waves. This transmitted energy is captured by the receiver and is converted into electric energy. This conversion is generally done by the piezoelectric effect [25–26]. Figure 1.5 represents the schematic of a basic AEH system. It consists of a resonator followed by a piezoelectric crystal and an energy storage device such as a battery. The sound energy amplification is done by the resonator, and then the piezoelectric crystal converts the sound pressure into electrical energy. In [27], for low-frequency bands, an AEH system is proposed which has extremely high efficiency. It is composed of an affixed piezo disk with tuneable Helmholtz resonator. This harvester can generate 3.491 W power with 100 dB sound pressure, and its conversion efficiency is as much as 38.4%.

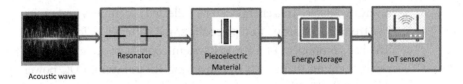

FIGURE 1.5    Schematic of AEH system

(v)    *Thermal EH*: Thermal EH is the method of harnessing power from thermal gradients using TEGs [28]. A thermoelectric harvester harnesses energy via the Seebeck effect, which asserts that when two dissimilar metals linked at two junctions are kept at differing temperatures, electrical voltage is created. This is due to the fact that various metals react differently to temperature differences, causing heat to flow through the TEG. As a result, the voltage difference between the hot and cold plates is proportional to the temperature differential. In [29], the authors have implemented TEG to power the wireless sensor nodes of a network. Similarly, IoT sensors are powered via thermoelectric harvesting in [30], providing an efficiency of 26.43%.

(B)    *Mechanical EH*: Mechanical vibrations and pressure are abundant energy sources that may be exploited to create sustainable IoT devices for a variety of applications.

(i)    *Piezoelectric EH*: Mechanical vibrations and pressure have sufficiently high energy that can be harvested using a piezoelectric harvester. It is based on the piezoelectric effect that refers to conversion of mechanical energy such as pressure, vibrations, or force into electrical power by straining a piezoelectric material. Applying pressure on the piezoelectric material, in particular, causes charge separation across the harvester, resulting in an electric field and thus voltage proportionate to the stress [31–32]. A major benefit of piezoelectric EH is that it does not need external voltage sources, has few movable parts, and can create power at easily adjustable voltage levels. Harvesters of this type have the advantages of high power density and simplified design. They utilise a wide variety of frequencies in their design and production. To show its efficiency, a test is presented in [33] where eight such harvesters are installed in a highway side area. The results have proved that harvesters are able to provide enough power to operate 24 light emitting diode indicators that ensure drivers' safety and monitor real-time traffic data.

(C) *Human Body-Based EH*: Energy is produced from the human body in a variety of ways, including finger motions (FM), footfalls (FF), breathing (Br), blood pressure (BP), and so on. This energy can be exploited specifically by wearable and implantable wireless devices, which are basically IoT devices that are used to monitor body conditions [34]. Human heat EH [35] is based on temperature changes in the body and employs two types of energy harvesters, thermoelectric and pyroelectric, working based on spatial temperature gradients and temporal temperature variation, respectively. The kinetic energy generated by the motion of body parts, termed biomechanical energy, can also be harvested. A report is presented in [36] showing the amount of power generated during the motion of body parts. When typing on a QWERTY keyboard, around 40–50 grams of figure pressure can produce up to 19 mW power. For footfall energy, a person of 68 kg in weight produces 324 W power walking two footsteps per second and around 1 watt power during breathing. These harvested energies can be utilised to self-power IoT sensors implanted in the body by eliminating the need for a battery. Tables 1.2 and 1.3 provide a comparative characterisation of ambient and human body energy-harvesting resources.

(D) *Hybrid EH*: There are several cases where harvesting power from a single energy source is not sufficient to power IoT sensors. For instance, the reliability of solar power is low, as it is affected by the weather conditions and temporal variations. Hybrid energy-harvesting technology is gaining traction as a solution to a single energy harvester's energy shortage. So, hybrid harvesting schemes using other supplies to complement the system are proposed in [37] and [38]. In general, this refers not only to harnessing energy from various sources but also converting energy into electricity using various types of transduction techniques with hybrid materials and mechanisms. Because ambient energy sources have a wide range of energy densities, hybrid EH systems are predicted to play a significant part in the IoT sector [39] and possess great potential to increase the lifetime of IoT devices by operating in complementary manner. Hybrid EH combining TEG with solar energy is presented in [40]. Hybrid EH is used to power the on-body sensor using RF energy harvesting and TEG to harvest human thermal energy in [41]. Although hybrid EH is still in its early stage of realisation, it is expected to play a key role to power IoT devices for smart cities [42]. The study has proved that this concept is one of the best to be applied in a hostile environment.

TABLE 1.2 Characterisation of Various Ambient EH Technologies

| S. No | Energy Source | Energy Scavenging Device | Power Density | Characteristics | References |
|---|---|---|---|---|---|
| 1 | Solar energy (outdoor) | PV cells | 15–100 mW/cm$^3$ (sunny day) 0.15–100 µW/cm$^3$ (cloudy day) | Uncontrollable, predictable, ambient | [43–44] |
| 2 | Solar energy (indoor) | PV cells | 6 µW/cm$^3$ | Non-ambient, controllable | [45–46] |
| 3 | RF | Rectenna | 0.08 nW–0.1 µW/cm$^3$ (GSM 900/1800MHz) 0.01 µW/cm$^3$ (Wi-Fi) | Predictable, available anywhere | [47] |
| 4 | Wind/air flow | Triboelectric generator | 400 µW/cm$^3$ | Uncontrollable, unpredictable | [42] |
| 5 | Mechanical energy/vibration/motion | Piezoelectric Electrostatic | 0.2 mW/cm$^3$ 184 µW/cm$^3$ | Non-ambient, unpredictable, controllable | [43, 45] |
| 6 | Thermoelectric | Seebeck effect | 40 µW–10 mW/cm$^3$ (5–20°C temperature gradient) | Ambient, uncontrollable, unpredictable | [47–48] |
| 7 | Acoustic noise | Diaphragms and piezoelectric transducers | 0.003 µW/cm$^3$ from 75 dB and 0.96 µW/cm$^3$ from 100 dB | Ambient, uncontrollable, unpredictable | [49–50] |

TABLE 1.3   Summary of Power Generated from Human Body Sources

| S. No. | Human Body–Based Source | Available Power | Usable Power |
| --- | --- | --- | --- |
| 1 | Finger motion | 6.9–19 mW | 0.76–2.1 mW |
| 2 | Footfall | 67 W | 5.0–8.3 W |
| 3 | Breathing | 0.83 W | 0.42 W |
| 4 | Blood pressure | 0.93 W | 0.37 W |

## 1.5  POWER MANAGEMENT IN IoT DEVICES

To enable long life for IoT devices, power utilisation by IoT devices is required to be reduced by selecting an efficient power management integrated circuit (PMIC). Several limitations must be considered while designing a PMIC, including the selection of transducer, anticipated power calculations based on generating values, and power consumption. Several PMICs are already devised that can provide an integrated solution for EH in IoT applications. Some of them that harvest solar energy are discussed in the following.

Because of depletion in conventional energy sources, many EH processes have been unleashed. With the advancement of IoT networks as the number of communicating IoT devices is increasing, additional power generation is also desirable. However, there are several challenges that should be highlighted in implementing EH systems for the IoT. For example, it is challenging for IoT devices to transmit or distribute scavenged energy, as ambient energy sources mostly have low power efficiency. Therefore, to enhance EH efficiency, power management of the harvested energy is an extremely challenging research component. Controlling power transmission from diverse power sources to IoT nodes is difficult and is frequently fraught with problems. The design of the power management unit must be such that it can route power depending upon the IoT application. One energy source is insufficient for relieving the current energy demand since it cannot give enough power to the sensors. As a result, multisource or hybrid EH systems are now being developed. These systems, however, necessitate a great deal of attention because they are difficult to install.

From an implementation point of view, the additional expenditure is an obstacle for mass production of harvesting modules; also, the power needs of some wireless technologies may not be met by harvesting technologies. It has been noticed that some EH units have worked proficiently for the last 15 years without hardware degradation [57], which shows regular

TABLE 1.4   PMIC Modules Currently Used in IoT Applications

| S. No | PMIC modules | Characteristics | Application |
|---|---|---|---|
| 1 | E-PEAS (AEM 10940) [52] | Uses PV cells to harvest energy up to 50 mW | Wearable devices, home automation, geolocation |
| 2 | Cypress (S6AE10xA) [53] | Harvests power using PV cells as small as 1 cm$^2$ | Automated buildings and smart home monitoring, agriculture and industrial sectors |
| 3 | Texas Instruments (Bq251120) [54] | High integration, automatically shuts off switches to reduce battery drainage | Wearable devices |
| 4 | Maxim Integrated (MAX14720) [55] | Highly energy efficient | Fitness and medical devices |
| 5 | STMicroelectronics (SPV1050) [56] | Highly efficient for the management of power, as can operate up to 400 mW | Smart homes, fitness, and wearable devices |

hardware maintenance is not required. The advancement of low-power electronic technologies, combined with cloud computing for data acquisition, cuts electrical energy usage even more. A high element of power saving can be in smart building systems where copper wires, materials, setup, and maintenance expenses can be decreased by EH strategies.

## 1.6 CONCLUSION

Harvesting energy plays a vital role in enhancing the efficiency and longevity of IoT devices. This technique is attracting tremendous attention from various research communities and the industries involved in the design and deployment of self-powered IoT devices. Various energy-harvesting techniques are discussed in this chapter, and also a comparative study and their applications are provided. With the implementation of EH to power IoT devices, problems such as battery replacement, specifically in inaccessible locations, can be eliminated and the devices can be made truly wireless without physical interventions. Also, they can be operated over the long term without the requirement of major maintenance. To reduce power consumption, power management integrated circuits play a vital role, which helps to enhance the lifespan of the IoT system. Power management integrated circuits that manage the power of IoT devices are also discussed. However, the practical implementation of EH systems is more complex than that of conventional battery-powered systems. The benefits of the removal of the battery and reduction of maintenance costs can outweigh the complexity issue by making IoT devices self sustainable. Hence, it is worth mentioning that EH has solid potential to come up with a long-lasting IoT network system.

## REFERENCES

[1] Jorge E. Ibarra-Esquer et al., "Tracking the evolution of the Internet of Things concept across different application domains," *Sensors*, vol. 17, no. 1379, pp. 1–24, 2017.
[2] F. Wang, L. Hu, J. Zhou and K. Zhao, "A survey from the perspective of evolutionary process in the Internet of Things," *International Journal of Distributed Sensor Networks*, vol. 2015, pp. 1–9, 2015.
[3] ITU Internet Reports 2005, *The Internet of Things, Executive Summary, International Telecommunication Union (ITU)*, Geneva, Switzerland, November 2005.
[4] A. Haroon et al., "Constraints in the IoT: The world in 2020 and beyond," *International Journal of Advanced Computer Science and Applications*, vol. 7, no. 1, pp. 252–271, 2016.
[5] K. Z. Panatik et al., "Energy harvesting in wireless sensor networks: A survey," in *Proceedings of IEEE 3rd International Symposium on Telecommunication Technologies (ISTT)*, pp. 53–58, Kuala Lumpur, Malasiya, 2016.
[6] S. Ma et al., "Temperature effect and thermal impact in lithium-ion batteries: A review," *Progress in Natural Science: Materials International*, vol. 28, no. 6, pp. 653–666, 2018.
[7] M.-L. Pykälä, K. Sipilä, U.-M. Mroueh, M. Wahlström, H. Huovila, T. Tynell and J. S. Tervo, "Energy harvesting roadmap for societal applications," in *VTT Research Report VTT*, Espoo, Finland, 2012.
[8] B. Franciscatto, "Design and implementation of a new low-power consumption DSRC transponder," PhD Thesis, Université de Grenoble, Saint-Martin-d'Hères, France, 2014.
[9] M. Shirvanimoghaddam et al., "Towards a green and self-powered Internet of Things using piezoelectric energy harvesting," *IEEE Access*, vol. 7, pp. 94533–94556, 2019.
[10] www.psma.com/HTML/newsletter/Q2_2012/page8.html, [Online].
[11] F. K. Shaikh and S. Zeadally, "Energy harvesting in wireless sensor networks: A comprehensive review," *Renewable and Sustainable Energy Reviews*, vol. 155, pp. 1041–1054, 2016.
[12] S. Sudevalayam and P. Kulkarni, "Energy harvesting sensor nodes: Survey and implications," *IEEE Communications Surveys & Tutorials*, vol. 13, no. 3, pp. 443–461, 2011.
[13] P. Luo, D. Peng, Y. Wang and X. Zheng, "Review of solar energy harvesting for IoT applications," in *Proceedings of IEEE Asia Pacific Conference on Circuits and Systems (APCCAS)*, pp. 512–515, Chengdu, China, 2018.

[14] P. Breeze, "Solar power," in *Power Generation Technologies*, 3rd edition, pp. 293–321, Newnes, UK and Oxford, 2019.

[15] T. Sanislav, G. D. Mois, S. Zeadally and S. C. Folea, "Energy harvesting techniques for Internet of Things (IoT)," *IEEE Access*, vol. 9, pp. 39530–39549, 2021.

[16] P. Sikka, P. Corke, L. Overs, P. Valencia and T. Wark, "Fleck—A platform for real-world outdoor sensor networks," in *Proceedings of 3rd International Conference on Intelligent Sensors, Sensor Networks and Information*, pp. 709–714, Melbourn, Australia, 2007.

[17] F. Simjee and P. S. Chou, "Everlast: Long-life, supercapacitor-operated wireless sensor node," in *Proceedings of the 2006 International Symposium on Low Power Electronics and Design*, pp. 197–202, Bavaria, Germany, 2006.

[18] M. Minami, T. Morito, H. Morikawa and T. Aoyama, "Solar Biscuit: A battery-less wireless sensor network system for environmental monitoring applications," in *Proceedings of 2nd International Workshop on Networked Sensing Systems; Association for Computing Machinery*, New York, NY, 2005.

[19] J. Polastre, R. Szewczyk and D. T. Culler, "Enabling ultra-low power wireless research," in *Proceedings of the IPSN 2005. Fourth International Symposium on Information Processing in Sensor Networks*, pp. 364–369, Boise, ID, 2005.

[20] Le-G Tran, H.-K Cha and W.-T Park, "RF power harvesting: A review on designing methodologies and applications," *Micro and Nano Systems Letters*, vol. 5, no. 1, pp. 1–16, 2017.

[21] M. Cansiz, D. Altinel and G. K. Kurt, "Efficiency in RF energy harvesting," *Energy*, vol. 74, pp. 292–309, 2019.

[22] C. Lin, C. Chiu and J. Gong, "A wearable rectenna to harvest low-power RF energy for wireless healthcare applications," in *11th International Congress Image Signal Process., Biomedical Engineering Informatation (CISP-BMEI)*, pp. 1–5, Beijing, China, 2018.

[23] S. Orrego et al., "Harvesting ambient wind energy with an inverted piezoelectric flag," *Applied Energies*, vol. 194, pp. 212–222, 2017.

[24] R. K. Sathiendran, R. R. Sekaran, B. Chandar and B. S. A. G. Prasad, "Wind energy harvesting system powered wireless sensor networks for structural health monitoring," *International Journal of Engineering Research & Technology (IJERT) NCACCT*, vol. 2, no. 12, pp. 523–526, 2014.

[25] J. Choi, I. Jung and C. Kang, "A brief review of sound energy harvesting," *Nano Energy*, vol. 56, pp. 169–183, 2018.

[26] M. Yuan, Z. Cao, J. Luo and X. Chou, "Recent developments of acoustic energy harvesting: A review," *Micromachines*, vol. 48, no. 10, pp. 1–21, 2019.

[27] M. Yuan, Z. Cao, J. Luo, J. Zhang and C. Chang, "An efficient low-frequency acoustic energy harvester," *Sensors and Actuators A: Physical*, vol. 264, pp. 84–89, 2017.

[28] K. Kobbekaduwa and N. D. Subasinghe, "Modelling and analysis of thermoelectric generation of materials using Matlab/Simulink," *International Journal of Energy and Power Engineering*, vol. 97, no. 5, pp. 97–104, 2016.

[29] H. Park et al., "Energy harvesting using thermoelectricity for IoT (Internet of Things) and E-skin sensors," *Journal of Physics Energy*, vol. 1, no. 4, pp. 1–15, 2019.

[30] D. Charris et al., "Thermoelectric energy harvesting scheme with passive cooling for outdoor IoT sensors," *Energies*, vol. 13, no. 11:2782, pp. 1–25, 2020.

[31] P. Jiao, K. Egbe, Y. Xie, A. Nazar and A. Alavi, "Piezoelectric sensing techniques in structural health monitoring: A state-of-the-art review," *Sensors*, vol. 20, no. 3730, 2020.

[32] A. G. Corina Covaci, "Energy harvesting with piezoelectric materials for IoT—Review," *Proceedings of ITM Web of Conferences*, vol. 29, no. 03010, pp. 1–15, 2019.

[33] J. Y. Cho et al., "A multifunctional road-compatible piezoelectric energy harvester for autonomous driver-assist LED indicators with a self-monitoring system," *Applied Energy*, vol. 242, pp. 294–301, 2019.

[34] B. Latré et al., "A survey on wireless body area networks," *Wireless Netw*, vol. 17, pp. 1–18, 2011.

[35] D. C. Hoang, Y. K. Tan, H. B. Chng and S. K. Panda, "Thermal energy harvesting from human warmth for wireless body area network in medical healthcare system," in *International Conference on Power Electronics and Drive Systems (PEDS)*, Taipei, Taiwan, pp. 1277–1282, 2009.

[36] T. Starner, "Human-powered wearable computing," *IBM Systems Journal*, vol. 35, no. 3.4, pp. 618–629, 1996.

[37] Q. Shi, Y. Yang, F. Wen, Z. Zhang, C. Lee and B. Dong, "Technology evolution from self-powered sensors to AIoT enabled smart homes," *Nano Energy*, vol. 79, article no. 105414, 2021.

[38] C. Wang, J. Li, Y. Yang and F. Ye, "Combining solar energy harvesting with wireless charging for hybrid wireless sensor networks," *IEEE Transactions on Mobile Computing*, vol. 17, no. 3, pp. 560–576, 2018.

[39] S. Chandrasekaran et al., "Micro-scale to nano-scale generators for energy harvesting: Self powered piezoelectric, triboelectric and hybrid devices," *Physics Reports*, vol. 792, pp. 1–33, 2019.

[40] J. Pradeep et al., "Hybrid energy harvesting system using IOT," *2020 IOP Conference Series: Materials Science and Engineering*, vol. 923, no. 012077, pp. 1–8, 2020.

[41] O. A. Sarareh et al., "A hybrid energy harvesting design for on-body Internet-of-Things (IoT) networks," *Sensors*, vol. 20, no. 2:407, pp. 1–16, 2020.

[42] O. Akan, O. Cetinkaya, C. Koca and M. Ozger, "Internet of hybrid energy harvesting things," *IEEE Internet of Things Journal*, vol. 5, no. 2, pp. 736–746, 2018.

[43] G. Zhou, L. Huang, W. Li and Z. Zhu, "Harvesting ambient environmental energy for wireless sensor networks: A survey,"*Journal of Sensors*, vol. 2014, article 815467, pp. 1–20, 2014.

[44] R. Jhonny, R. William and M. S. Litz, "Low light illumination study on commercially available homojunction photovoltaic cells," *Applied Energy*, vol. 191, no. 1, pp. 10–21, 2017.

[45] N. Garg and R. Garg, "Energy harvesting in IoT devices: A survey," in *International Conference on Intelligent Sustainable Systems (ICISS)*, pp. 127–131, Palladam, India, 2017.

[46] F. De Rossi, T. Pontecorvo and T. M. Brown, "Characterization of photovoltaic devices for indoor light harvesting and customization of flexible dye solar cells to deliver superior efficiency under artificial lighting," *Applied Energy*, vol. 156, no. 1, pp. 413–422, 2015.

[47] S. Basagni, M. Y. Naderi, C. Petrioli and D. Spenza, "Wireless sensor networks with energy harvesting," in *Mobile Ad Hoc Networking:Cutting Edge Directions*, 2nd edition, Wiley, NY, USA, pp. 1–36, 2013.

[48] M. Stordeur and I. Stark, "Low power thermoelectric generator—self sufficient energy supply for micro systems," in *Proceedings of 16th International Conference on Thermoelectrics*, IEEE, pp. 575–577, Dresden, Germany, 1997.

[49] L. H. Fang, S. I. S. Hassan, R. Bin Abd Rahim and J. M. Nordin, "A review of techniques design acoustic energy harvesting," in *Proceedings of IEEE Student Conference on Research andDevelopment (SCOReD)*, pp. 37–42, Kuala Lumpur, Malaysia, 2015.

[50] J. Choi, I. Jung and C.-Y. Kang, "A brief review of sound energy harvesting," *Nano Energy*, vol. 56, pp. 169–183, 2019.

[51] M. K. Mishu et al., "Prospective efficient ambient energy harvesting sources for IoT-equipped sensor applications," *Electronics*, vol. 9, no. 9, pp. 1–22, 2019.

[52] F. Zhang et al., "A Batteryless mW MICS/ISM-band energy harvesting body sensor node SoC for ExG applications," *IEEE Journal Solid-State Circuits*, vol. 48, pp. 199–213, 2012.

[53] H. Sharma, A. Haque and Z. Jaffery, "Solar energy harvesting wireless sensor network nodes: A survey," *Journal of Renewable and Sustainable Energy*, vol. 10, no. 2, article 023704, 2018.

[54] Tech-Edge: TI Introduces the Industry's Lowest Power Battery Management Solution for Wearable and IoT. www.tech3dge.com/ti-introduces-the-industryssmallest-and-lowest.

[55] J. Wilden, A. Chandrakar, A. Ashok and N. Prasad, "IoT based wearable smart insole," in *2017 Global Wireless Summit (GWS)*, pp. 186–192, Cape Town, SA, 2017.

[56] K. Kadirvel et al., "A 330nA energy-harvesting charger with battery management for solar and thermoelectric energy harvesting," in *Proceedings of IEEE International Solid-State Circuits Conference*, pp. 106–108, San Francisco, 2012.

[57] S. Chiriac and B. Rosales, "An ambient assisted living monitoring system for activity recognition—Results from the first evaluation stages," in *Ambient Assisted Living-Kongress*, Berlin, Germany, pp. 135–146, 2012.

# CHAPTER 2

# Efficient Ambient Energy-Harvesting Sources with Potential for IoT and Wireless Sensor Network Applications

Sonam Lata, Shabana Mehfuz

## CONTENTS

DOI: 10.1201/9781003218760-2

## 2.1 INTRODUCTION

The technique of obtaining energy from the environment is known as ambient energy harvesting. Solar and wind power, ocean waves, piezo-electricity, thermoelectricity, and physical movements are among the options available for energy scavenging. Recent investigations on prospective ambient energy-harvesting sources and systems for the Internet of Things (IoT) and wireless sensor networks (WSNs) are examined in this chapter. The ability to meet the power needs of autonomous wireless and portable devices is a major concern today. Energy storage has considerably improved in recent years. Battery-powered sensor nodes and components cannot work for a long duration. Energy conservation might be problematic in a large network with lots of sensor devices. So in terms of reducing maintenance and operational expenses, ambient power sources are being examined as a backup to batteries. Energy harvesting is the process of converting ambient energy into usable electrical energy. When compared to energy stored in conventional energy storage systems such as batteries and capacitors, the environment provides a virtually endless supply of useful energy. As systems shrink in size, less power becomes accessible, leading to a restricted run-time for batteries.

This chapter examines the benefits and drawbacks of energy harvesting–based WSNs as well as why standard WSN protocol stacks must be completely redesigned. We begin by discussing an energy harvesting–based WSN node's architecture, particularly its energy subsystem. Various types of energy-accessible methods for harnessing them are then presented. Models for estimating wind and solar energy availability are also explained. After that, we look at the job allocation, media access control (MAC), and routing protocols that have been suggested so far for energy harvesting–based WSNs.

The Internet of Things and WSNs have also received a lot of attention in recent years as a result of their widespread implementation in meeting worldwide demand. The decreased lifespan of different energy supplies required to power sensors over time limits the attraction of sensor and low-power digital devices. Models for wind and sunlight harvesting have also been investigated in order to improve the reliability of the power output required by IoT devices. This chapter also examines the challenges that may be overcome in order to make IoT-enabled sensors more durable, reliable, energy efficient, and cost efficient.

## 2.2  AMBIENT ENERGY SOURCES

Ambient energy harvesting is the technique of gathering and converting energy from the environment and storing it for use in electronics applications. Solar energy, ocean waves, piezoelectricity, thermoelectricity, and physiological motions are all potential sources of energy.

Prolonged product life is essential; it also has larger benefits in systems with strong encryption, like those used to supervise a machine in a production facility. The long-term remedy must be irrespective of the limited energy available when certain devices are active or operating. Table 2.1 and Figure 2.1 compare the estimated power and challenges of several environmental energy sources.

EH processes from various environmental sources are compared in Table 2.1. This analysis is divided into the following sections: fundamental ambient supplies, properties, specific energy, conversion ratio, energy handiness, harvesters, benefits, and constraints. Sunlight, air, heat

TABLE 2.1  Power Density for Different Energy-Harvesting Sources

| Supplier of Power | Concentration of Power |
|---|---|
| Acoustic Noise | $0.003 \ \mu W/cm^3$ |
| | $0.96 \ \mu W/cm^3$ |
| Temperature Variation | $10 \ \mu W/cm^3$ |
| Ambient Radio Frequency | $1 \ \mu W/cm^2$ |
| Thermoelectric | $60 \ W/cm^2$ |
| Vibration | $4 \ W/cm^3$ (human motion—Hz) |
| (Micro-Generator) | $800 \ W/cm^3$ (machines—kHz) |
| Vibrations (Piezoelectric) | $200 \ \mu W/cm^3$ |
| Airflow | $1 \ \mu W/cm^2$ |
| Push Buttons | $50 \ J/N$ |
| Shoe Inserts | $330 \ \mu W/cm^2$ |
| Hand Generators | $30 \ W/kg$ |
| Heel Strike | $7 \ W/cm^2$ |
| Photovoltaic | Outdoors (direct sun): $15 \ mW/cm^2$ |
| | Outdoors (cloudy day): $0.15 \ mW/cm^2$ |
| | Indoors: $<10 \ W/cm^2$ |
| Sunlight | $10–100 \ mW/cm^2$ |
| RF | $1 \ \mu W/cm^2$ |
| Exhalation | $0.4 \ W$ |
| Fuel Cell | $10 \ \mu W$-$2 \ mW$ |
| E-Field | $17 \ \mu W/cm^3$ |
| M-Field | $150 \ \mu W/cm^3$ |
| Breathing | $0.42 \ W$ |

energy, movement, movement, radio frequency (RF), airflow, magnetic field (M-field), and electric field are examples of natural ambient sources (E-field). Breathing (Br.), exhalation (Ex.), finger motion (FM), footfalls (FF), and blood pressure (BP) are examples of human-dependent ambient energy sources.

A piezoelectric generator generates $200 \ W/cm^2$ energy from various movements. The energy generated from motion is partially predictable and controlled. A thermocouple collects $60 \ W/cm^2$ of electricity from thermal energy sources at $DT = 5 \ C$ with an efficiency of 1% when DT is 40%. Thermal energy is both ambient and predictable, yet it is unregulated. Electromagnetic induction sources use vibrational energy sources to create $0.2 \ mW/cm^2$ of energy. By employing piezo turbines and an anemometer, airflow generates $100 \ mW/cm^2$ energy. Establishing an ambient airflow is unregulated, time-consuming, and inconsistent. But by using a piezoelectric energy harvester, finger movements or footsteps may also be revealed to be a controlled sort of power source, producing 2.1 mW and 5 W energy, respectively. The FM source has a conversion efficiency (CE) of 11%, while the FF source has a CE of 7.5%. To gather energy from breathing movements, a ratchet-flywheel (R-F) with a 50% CE might be used. It emits 0.42 W of power. The quantity of inhaling energy gathered is erratic and unpredictable. A current transformer (CT) is a method for extracting energy by using a magnetic field. This form of energy is predictable and controlled, with a power output of 150 $W/cm^3$.

The application and extension of the M-field energy harvester are limited by a higher current demand. Furthermore, when a metallic plate energy harvester is utilised, an electric field creates 17 $W/cm^3$ energy. When paired with a micro-generator and a breath mask, human-based attributes such as blood pressure and expiration create 40% more energy. Exhalation may yield 0.4 W of energy, while BP generates 0.37 W. The microbial fuel cell (MFC), which can generate 10 W–2 mW of energy, is extensively used in biosensors. This chapter presents a wide range of potential approaches for collecting and storing energy from a variety of ambient energy sources. Thermal energy, on the other hand, is limited by the relatively minor temperature fluctuations that occur throughout a chip. Although vibration energy is a low-energy source, it is suited for applications [1].

Ambient energy may be derived from biological and chemical substances, as well as radiation. A notional block diagram of general ambient energy-harvesting devices is shown in Figure 2.2. The first sequence highlights the sources of power. The second sequence indicates the successful execution and technologies used to extract power from the source. The third sequence displays every source's energy-harvesting approaches. Several authors conducted an extensive study of the literature on prospective energy-scavenging approaches. The findings of this literature review have been organised by source and are presented in the study's parts.

## 2.2.1 Mechanical Energy Harvesting

The self-powered, rechargeable, portable wheeled computer mouse was presented as a demonstration of generating electricity by rotating action by the authors of [2]. The device was distinctive in that it used the cursor disc to capture vibrational mobility in terms of generating and obtaining electric power. The electric generator was triggered by pushing the mouse to generate rolling energy. The mouse only needs 2.2 mW of power to work. A power generation apparatus produced more than 3 mW of energy, which would have been sufficient for cordless mouse functioning over a 1-meter transmittal range.

The authors of [3] provided another demonstration of mechanical energy harvesting in the form of an electret-based electrostatic micro-generator. A micro-machined electrostatic converter was used in a system together with a vibration-reactive variable capacitor polarised by an electret. In the same study, a global multi-domain model was established and examined, revealing that output power potential of up to 50 watts for a 0.1 cm² surface area was possible.

## 2.2.2 Mechanical Vibrations

Internal machinery sensors may contain significant amounts of structural vibrations that can be consistently documented and then used. There are electromechanical or piezoelectric acoustic emission energy recovery technologies. Electromechanical harvesting techniques, on the other hand, have received greater attention and use. The movement of a spring-mounted entity surrounding its supporting structure could have an influence on the separation of electrical generation from sound waves [4]. As a result of vibrations, the mass component moves and oscillates, resulting in mechanical acceleration.

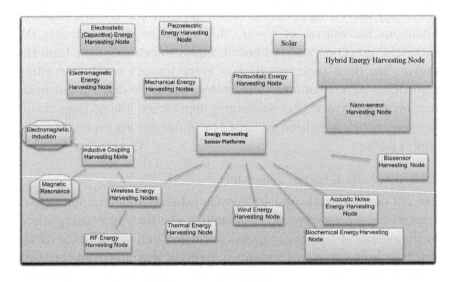

FIGURE 2.1 Different energy types and sources

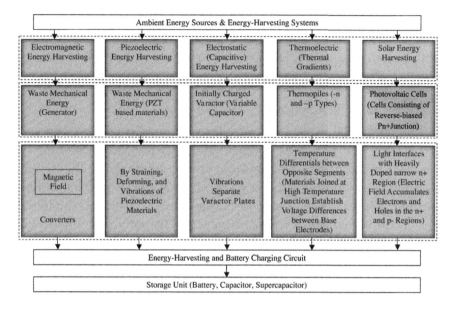

FIGURE 2.2    Ambient energy systems

### 2.2.3  Electromagnetic

Using a magnetic field, this technology converts mechanical energy to electrical energy [5]. A wire linked to the oscillating object is designed to generate electricity by flowing through a motionless magnet's magnetic field. According to Faraday's law, the coil creates a voltage by passing through a changing quantity of magnetic flux. Because induced voltage is intrinsically low, it must be raised to be a viable kind of power. It is possible to raise the electric potential by using a transformer, expanding the amount of turns on the coil, or raising the stable magnetic field [1]. The size and material features of the microchip limit each of these attributes.

### 2.2.4  Piezoelectric

By rupturing a piezoelectric material, mechanical energy can be converted to electrical energy [6]. Limited voltage-based applications vary in time and produce a discontinuous AC supply [7]. When compared to electromagnetic energy transformation, piezoelectric conversion of energy yields higher voltage and power density ranges. Furthermore, piezoelectricity denotes a material's ability to generate an electric potential in addition to mechanical stress, such as crystals and some forms of ceramics [8]. If the piezoelectric material is not short-circuited, the applied mechanical stress generates a voltage across it. Piezoelectric materials are used in a multitude of scenarios, including electric cigarette lighters. Portable systems with built-in piezoelectric ignition systems are used to start gas barbecues, stoves, and various gas burners.

## 2.2.5 Electrostatic (Capacitive)

The authors in [9] worked on capitalising the modifying capacitance of sound varactors. To convert mechanical energy into electrical energy, a variable capacitor uses sound waves to distinguish its plates. The conversion was done utilising either a constant voltage or a constant current method. When a variable capacitor's capacitance fluctuates after an initial charge, the voltage across it remains constant. The authors of [10] investigated and evaluated three major resonance renewable resources (electrostatic, electromagnetic, and piezoelectric) by comparing based on their volatility, available power, size, and challenges in a study. Table 2.2 summarises the investigation's findings.

## 2.2.6 Thermal or Thermoelectric Energy Harvesting

Thermal gradients in the environment, according to [11–12], are quickly converted to electrical energy via the Seeback (thermoelectric) effect (1999). Temperature fluctuations among opposing sectors of a metallic conductor can create heat transfer and charge flow because mobile, increased energy components disperse from high to low concentration locations. To achieve acceptable voltage and power levels, greater temperature gradients are required [7]. Temperature variations of more than 10°C are unusual in microsystems, resulting in low voltage and power levels. Furthermore, naturally occurring temperature variations can be used to scavenge energy from high-temperature situations. The thermoelectric generator (TEG) was created and implemented by Pacific Northwest National Laboratory [13], and it is one of the most modern types of thermoelectric energy harvesters. For a range of small applications, such a thermoelectric generator converts thermal energy from the external environment (ambient) into electrical power. A thermoelectric energy generator might be used for a variety of simple and stand-alone applications.

A thermoelectric generator was created and displayed by Applied Digital Solutions Company. This thermoelectric generator, which is 0.5 cm$^2$ in size and a few centimetres thick, can generate 40 mw of electricity from temperature variations of 5°C [14]. This device's output voltage is roughly 1 V, making it ideal for good electrical applications. The thermal-expansion-actuated piezo-electric power supply has indeed been proposed as a means of energy conversion from temperature variation in the environment [15, 16].

### 2.2.7 Light Energy (Solar Energy) Harvesting

Light energy is converted into electrical energy by a photovoltaic cell [17]. Light travels via the highly conservative and limited n+ region of each cell, which is made up of a reverse biased pn+ junction. When a circuit is connected, the collection of electrons are emitted through the load and recombine with holes on the p-side, producing a photocurrent that is proportional to the amount of light and independent of cell voltage. Several previous research projects have shown that photovoltaic cells can power a microsystem. Furthermore, by significantly increasing the exposed internal surface area of the device, a three-dimensional diode structure built on an absorbent silicon substrate improves efficiency [18].

### 2.2.8 Acoustic Noise

Compression waves obtained by an excitation frequency start causing acoustic noise. Pressure waves are detected by the human ear and converted into electrical impulses. Frequencies ranging from 20 Hz to 20 kHz may be detected by the human ear. Acoustic noise is made up of two components: acoustic power and acoustic pressure. The entire quantity of sound radiation emitted by a sound source during a certain time period is known as acoustic power. The hearing threshold of the human ear, which is fixed at 20 microPa, is used to assess acoustic pressure. The decibel, or Bel, is the unit of measurement used to quantify these relative sound levels (1 Bel equals 10 decibels) [19].

TABLE 2.2 Comparison of Vibration Energy-Harvesting Techniques

|  | Electrostatic | Electromagnetic | Piezoelectric |
|---|---|---|---|
| Complexity of process flow | Low | Very high | High |
| Energy density | 4 mJ cm$^3$ | 24.8 mJ cm$^3$ | 35.4 mJ cm$^3$ |
| Current size | Integrated | Macro | Macro |
| Problems | Very high voltage and need to add charge source | Very low output voltages | Low output voltages |

Occasional initiatives to absorb acoustic noise from a region with large and persistent sound levels in transforming it into electrical energy have been made. Acoustic energy conversion was researched by a team of researchers from the University of Florida. They used a flyback converter circuit to conduct a strain energy conversion experiment [20]. An AC to DC flyback converter was connected to the output of a vibrating piezoelectric transduction piezoceramic beam, with an effectiveness of more over 80% at 1 mW input power and 75% at 200 W input power [21].

### 2.2.9 Human Power

Cranking, shaking, squeezing, spinning, pushing, pumping, and tugging are just a few of the methods used to generate energy from active/passive human strength [22]. In the early twentieth century, certain spotlights were powered by wind-up generators.

A battery-free wireless remote control for Zenith televisions was another human-powered device. Robert Adler first revealed the concept, dubbed "Space Commander," in 1956. To generate ultrasonic waves, the device used a series of buttons pressed on aluminium material. The ultrasonic energy was deciphered by the television, allowing it to turn on, change stations, and mute the volume [23].

The authors in [24] developed a piezoelectric element with a resonantly tuned transformer and filtering circuitry. This device produced roughly 1 mJ at 3 V for every 15 N push whenever a trigger was pulled. The power generated was sufficient to power a digital encoder and a radio with a range of nearly 50 feet. Materials used for this device were off-the-shelf components, which enabled placing compact digital controllers independently without any battery or wire maintenance.

Several of these energy-harvesting technologies for powering wearable devices have been developed and tested by Starner [22]. Researchers arrived at the conclusion that the most predictable and sustainable form of energy emerges during heel strikes when jogging or walking [25]. Ongoing research focuses on moving energy from the shoe, where it is created, to the place of interest or use. Passive power sources are those that provide power without requiring the individual to do additional effort, such as walking or jogging.

There are several existing and underutilised principles for offering extra energy production to wireless electronic devices in power generation. The concept of total reliance on renewable sources of energy may aid in overcoming some of the constraints imposed by standard batteries' limited reliability. By collecting ambient energy, conventional electronic part lifespan and support could be extended.

## 2.3 ROLE OF ENERGY HARVESTING IN WIRELESS SENSOR NETWORKS

WSN nodes are commonly powered by batteries. When a node's energy is expended, it dies. Batteries can only be replaced or recharged in extremely specific situations. Even if this is possible, the replacement/recharging method is time consuming and costly, as well as a performance drain on the network. As a result, many solutions for delaying battery depletion, such as power regulation and duty cycle-based operation, have been proposed. The latter method makes use of low-power wireless transceivers, whose components may be turned off to save energy. When the transceiver is turned on, the node's usage is significantly lower than when it is in low power mode (or "sleep"). While the node is sleeping, it is unable to broadcast or receive packets. The most popular type of solution for permitting long-lasting WSNs is to use protocols with extremely low duty cycles. Even if batteries aren't used very often, leaking depletes them in a matter of years.

Energy-harvesting WSNs have been created by giving WSN nodes the capacity to acquire energy from their exterior. Solar power, wind, mechanical vibrations, temperature extremes, magnetic fields, and other energy sources may be used in energy harvesting. WSN nodes with energy-collecting subsystems can potentially last eternally by continually delivering energy and storing it for later use.

### 2.3.1 Node Platforms

Edge devices in energy harvesting-based WSNs may harvest energy in a diversity of ways and convert it into usable electrical power, in addition to sensing and communications systems.

### 2.3.2 Sensor Node Layout with Ability to Extract

A wireless sensor node's system design contains the following components (Figure 2.3):

- A power distribution module takes electrical energy from the harvester and either reserves it or delivers that to the other network elements for instant use;

- Energy storage, which permits produced energy to be saved for later use;

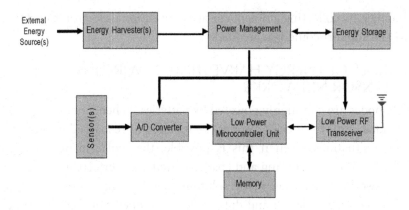

FIGURE 2.3   A wireless node's framework with energy storage devices

- A microcontroller is a device that allows you to control electronic devices;

- A data-transmitting and data-receiving radio transceiver;

- Sensory devices;

- An A/D converter to transform the analogue signal provided by the sensors into a digital signal that can be processed by the microcontroller; and

- Memory for storing detected data relating to the application and code.

The next section explores the energy recovery aspects of an energy harvesting–based WSN node, including various abstractions.

## 2.3.3 Harvesting Hardware Models

Figure 2.4 depicts the architectural configuration of a wireless sensor network base station battery system with accessible energy sources. The energy module consists of one or more harvesting processes that turn environmental energy into electrical energy. A more reasonable and logical design enables the node to directly use the collected energy and also involves a system that may function as the system's battery system, with the main objective of collecting and preserving the received energy [26].

Supplementary battery packs and super-capacitors are two common types of energy storage devices (also known as ultra-capacitors). They

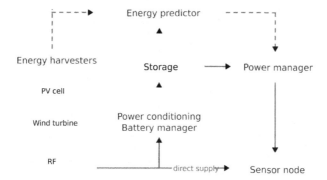

FIGURE 2.4 Basic structure of a wireless sensing node's power subsystem with energy storage abilities

may also be charged quickly utilising simple charging methods, reducing complexity and cost by eliminating the requirement for full-charge or deep-discharge protection circuits. They are also more efficient than electrochemical batteries in terms of recharging [27]. One benefit is that the atmospheric issues associated with battery dumping are no longer an issue. Super-capacitors are employed as energy storage in a variety of systems with harvesting capabilities, either alone [28] or in combination with batteries [29]. Other approaches, on the other hand, focus on platforms that are only powered by backup batteries [30].

In a variety of ways, both types of energy storage systems differ from perfect energy buffers: those that have a finite size $B^{Max}$ and will only keep a finite source of power, which have a charging efficiency $\eta c < 1$ and a discharging efficiency $\eta d < 1$, implying that some energy is dissipated while charging the battery, and those that tend to experience leaking and self-discharge, suggesting that some collected energy is dissipated even when the buffer is not being used. Electrochemical super-capacitors have a significant effect in the charged state. As the inter-plate voltage drops, they progressively lose energy. Numerous features enable the node to use the energy it has collected right away, reducing energy damage due to buffer problems.

## 2.3.4 Super-Capacitor Leakage Models

It's important to take leakage current into account when engaging with energy-harvesting devices, especially if the application scenario requires the acquired energy to be held for a significant duration. In general, if the energy source is unreliable, the amount of energy lost by leakage might be substantial. Super-capacitor leakage is affected by the capacitive reactance,

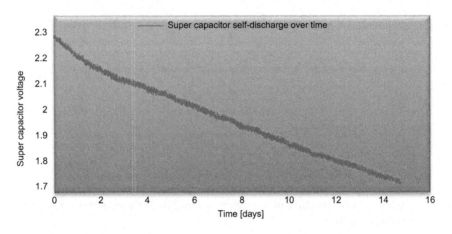

FIGURE 2.5    A super-capacitor's self-discharge with time

the quantity of energy deposited, the temperature range, the charge time, and other parameters. As a result, the leaking configuration of a super-capacitor is often computed [31]. Leakage from a charged super-capacitor has been proposed in the literature as a current, an increasing portion of the new super-capacitor voltage, a polynomial evaluation of its empirical leakage pattern, and eventually a linear estimation of its empirical leakage pattern [27, 31–32]. These models were created in response to super-capacitor leakage data, such as the two-week self-discharge of a charged 25F super-capacitor depicted in Figure 2.5.

Another thing to keep in mind in the super-capacitor vs. battery comparison is that the super-capacitor's entire energy cannot be used in many situations. The voltage of a super-capacitor declines linearly from maximum voltage to zero, unlike normal electrochemical batteries, which have a flat curve. The proportion of charge accessible to the sensor node is determined by the platform's voltage needs. For example, a Telos B mote requires a voltage range of 1.8 to 2.1 V. When the power supply of the super-capacitor starts falling beneath this threshold, the super-capacitor's remaining energy can no longer be used to power the node. This can be partially remedied by adding a DC-DC converter to broaden the voltage range but at the expense of its abilities and a primary source of energy consumption.

## 2.3.5  Battery Models

As they possess a specific range of energy units, batteries are energy storage technologies. They have charging and discharging efficiency strictly less than 1, which means that some energy is wasted while draining the battery. Batteries also have certain non-linear characteristics [33]. Furthermore, rechargeable batteries lose capacity with each recharge cycle, and their voltage is dependent on the battery's charging state and changes during discharge. When dimensioning and modelling energy-harvesting systems, these features should be taken into account because they can easily lead to inaccurate battery lifespan estimates. If somehow the power generated from the environment is retained in a battery bank, for example, it is essential to understand that the battery's capacity will be lowered with each recharging cycle, affecting the battery's provided capacity and lifespan.

In the recent literature, numerous different battery models have been published [33]. The following are some of them. (1) Models that replicate the physical processes and occur inside an electrochemical battery. These models are often extremely accurate, but they are computationally difficult and require a significant amount of configuration work. (2) Empirical models that use basic equations to estimate a battery's discharge behaviour. They are the ones that are not always right. They do, however, need a minimal amount of computing resources and configuration work [34]. Combined concepts, which provide a high-level description of a battery as well as explanatory expressions, are based on low-level evaluation and physical laws [35]; abstract models, such as stochastic systems [36]; electrical-circuit models [37]; and discrete-time very high speed integrated circuit hardware description language specification [38].

### 2.3.6 Prediction Models

Energy prediction techniques may be used to forecast source availability and estimate predicted energy consumption in the case of predictable, non-controllable power sources like solar [32]. Such a prediction model can solve the challenge of producing energy that is neither constant nor continuous, allowing management to make important decisions about using the energy it has. This subsection reviews the literature on diverse energy predictors for different sources of energy harvesters: wind and solar harvesters.

### 2.3.7 Exponentially Weighted Moving Average

Based on an exponentially weighted moving-average (EWMA) filter, Kansal et al. [32, 39] introduced a wind and solar power forecasting methodology. This concept is based on the assumption that the energy required at any given moment is the same. Time is split into $N$ equal lengths of 30-minute segments.

The quantity of energy accessible in former sessions usually kept as a weighted average, with older data contributing less. More formally, the EWMA model predicts that in time slot $n$ the amount of energy $\mu^{(d)} = \alpha \cdot x_n + (1 - \alpha) \cdot \mu^{(d-1)}$ will be available for harvesting, where $x_n$ is the amount of energy harvested by the end of the $n$th slot; $\mu^{(d-1)}$ is the average over the previous $d - 1$ days of the energy harvested in the $n$th slot; and $\alpha$ is a weighting factor, $0 \le \alpha \le 1$. The everyday solar energy cycle is used by EWMA to adjust to temperature fluctuations. In the case of small weather variations, the predicted estimates are fairly accurate. When wind patterns change often, EWMA does not adjust well to fluctuations in wind and solar patterns.

## 2.3.8 Weather-Conditioned Moving Average

The authors in [40] presented the prediction technique weather-conditioned moving average (WCMA) to solve the inadequacies of EWMA. Like EWMA, WCMA considers the energy accumulated during prior weeks. However, it takes into account the meteorological conditions of the current and preceding days. The D×N matrix **E**, where $D$ is the number of days considered and $N$ is the number of time slots each day, is stored in WCMA. On day $d$, the entry $E_{d,n}$ records the energy collected at time slot $n$. A vector **C** of dimension $N$ holds the energy of the current day. WCMA also preserves a vector **M** of size $N$, the $n$th item $M_n$, which contains the energy value observed during time slot $n$ in the previous $D$ days:

$$M_n = \frac{1}{D}\sum_{i=1}^{D} E_{d-i,n}$$

At the completion of each day, **M** is updated with the energy that was just measured, overwriting the previous day's date. $\bullet\ C_n + (1 - \alpha) \bullet M_{n+1} \bullet GAP_\alpha^K$ is the amount of energy predicted by WCMA for the next time slot $n + 1$ of the current day, where $C_n$ is the amount of energy observed during time slot n of the current day; $M_{n+1}$ is the average of the energy harvested during time slot $n + 1$ over the last $D$ days; and $GAP_\alpha^K$ is a rating factor that represents how the atmosphere has changed during time slot $n$ of the current day in comparison to the previous $D$ days. WCMA has been proven to have average prediction errors that are almost 20% lower than those of EWMA in the event of continuously evolving meteorological circumstances. Bergonzini et al. [41] developed a new version of WCMA. The researchers observed that the WCMA prediction error had discrete peaks at dawn and sunset, with values larger than 0.5 indicating a significant increase. This is because WCMA bases its energy projections on the value seen in the preceding slot. Because the solar conditions change dramatically at dawn and dusk, forecast errors are larger. To solve the problem, the authors proposed using a feedback mechanism known as a phase displacement regulator, which reduces the WCMA prediction error in a reasonable manner.

A predictor for ETH by Zurich's Moser et al. [42] presented a weighted sum of an historical data-based prediction approach. Solar power is assumed to be regular on a daily basis by the ETH prediction algorithm. As in previous systems, time is partitioned into fixed-length $T$ time slots. The power production during time slot $t$ is symbolised by $ES(t)$. The ETH predictor unit accepts the amount of energy collected $ES(t)$ for time $t \geq 1$ as an input and outputs $N$ future energy forecasts. All of the prediction intervals are of the same length, $L$, which is a multiple of $T$. $H = NL$ represents the whole forecast horizon. $P_S(t, k) = P_S(t + kL)$, $0 \leq k \leq N$ for the next $N$ prediction intervals are computed at each time slot for the following $t$ predictions about future energy availability $P_S(t, k)$. The contribution of older data diminishes exponentially, similarly to in EWMA. Noh and Kang [42] presented a solution that is comparable to prior techniques. They employed the EWMA equation to keep track of historical solar energy trends. They added a scaling factor $\Phi_n$ to alter future energy estimates in order to accommodate short-term fluctuating weather conditions. This factor is calculated as: $\Phi_n = \dfrac{x_{n-1}}{\mu_{n-1}}$, where $x_{n-1}$ represents the amount of energy collected by the end of slot $n - 1$, and $\mu_{n-1}$ represents the EWMA forecast of the amount of energy harvestable during slot $n - 1$. As a result, $\Phi_n$ is the difference between the actual collected energy and the projected energy for time slot $n$. This scaling factor is then used for subsequent projections to make them more accurate.

Pro-Energy (PROfile energy prediction model, Spenza and Petrioli [43]) is an energy prediction model for both solar and wind-based energy-harvested WSNs based on historical energy observations. Pro-basic Energy's idea is to use gathered profiles to show the energy accessible throughout various sorts of "average" days. Days, for example, might be described as bright, overcast, or rainy, with each variety having its own set of characteristics. Each day is divided into a fixed number of time slots ($N$). Predictions are made once for each position. In a $D \times N$ matrix $E$, a pool of previously

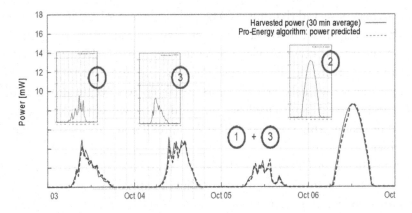

FIGURE 2.6    Pro-Energy predictions

observed energy profiles is also kept. The energy collected during a specific number $D$ of average days is represented by these profiles. By examining the stored profile that is the most comparable to the present day, Pro-Energy calculates the projected energy availability for the following time slot. The Euclidean spacing between vector fields is derived by considering the first $t$ constituents of the vectors and estimating the correlation between two independent profiles.

Figure 2.6 shows the Pro-Energy algorithm in operation over the course of 4 days of solar estimates. The first saved profile among the typical ones is picked during the early time slots of October 23rd (day 1), since it is the most comparable to the section of the present day seen thus far. The form of the profile varies throughout the day as fresh observations are made. In comparison to past systems, Pro-Energy performs admirably. Pro-Energy may significantly mitigate prediction errors, even in situations with a mixed bag of bright and gloomy days, a situation where EWMA fails miserably. We wrap up this part by discussing a method for predicting energy on medium-term timescales.

The researchers of [44] investigated a technology for solar and wind-powered edge devices that uses weather predictions to generate energy-harvesting estimations. The method was developed on the notion that calculating energy from past observations is more accurate than utilising weather forecasting data during medium timescales (3 hours to 3 days). The effectiveness of the suggested approach is calculated by evaluating its behaviour compared to that of primary power forecasters based on historical data.

### 2.3.9  Protocols for Energy Harvesting–Based Wireless Sensor Networks

In this part, we describe techniques for energy harvesting–based wireless sensor networks, with an emphasis on protocols from research fields that have gotten more attention, as well as work assignments to sensors and MAC and routing techniques.

### 2.3.10  Task Allocation

This section starts with a classification of activities based on their nature and characteristics, then moves on to task scheduling techniques. The following are the different types of tasks:

- Periodic vs. Aperiodic. Tasks are classified as periodic or aperiodic. Periodic jobs are scheduled to come at regular intervals and have a set arrival time. On-demand activities, also known as aperiodic tasks, have arbitrary arrival patterns.

- Preemptive vs. Non-preemptive. A preemptive active job can be preempted at any moment, but a non-preemptive activity can't be interrupted or stopped while it's running.

- Dependent vs. Independent. The execution of a job is described as independent if it is not reliant on the running or completion of other tasks. A dependent task cannot execute until the execution of certain other activities is complete.

- Node vs. Network Tasks. Each energy harvesting–based wireless sensor network's node has the ability to schedule both node and network duties. Node tasks include things like sensing, computing, and communication. Routing, leader election, cooperative communication, and other network functions are examples. Because node and network jobs have distinct properties, they require separate scheduling and energy budgeting methods.

The task's deadline is the date by which it must be finished. A deadline violation happens when the task deadline passes without the task being completed. The quantity of energy required to complete a job is known as its power demand.

Each job $T$ may have a value or reward $r$ connected with it, reflecting its importance. Task priority [45], invocation frequency [46], utility [47], or any other parameter can be used to determine rewards. If for job $i$, $T_i$ completes by the deadline, only then does it contribute $r_i$ units to the overall system reward. Each task's reward (priority) may shift over time. The pace at which the job runs is known as running speed. By dropping the processor's frequency response (CPU speed) and power consumption, dynamic voltage and frequency selection (DVFS) techniques may be employed to adjust running speed [45]. As the processor's operating frequency and voltage change, so does the task execution speed. Adjusting task pace is advantageous since it permits a node to change the job's runtime based on available forms of energy.

### 2.3.11 Protocols for Scheduling Node Jobs

One of the early works in energy harvesting–based WSN task scheduling is the lazy scheduling algorithm (LSA) [48]. Tasks are dynamically planned based on future energy availability, energy storage capacity, residual energy, and the sensor node's maximum power usage. The energy variability characterisation curve (EVCC), introduced by LSA, characterises the dynamics of the energy solution. There were several faults with LSA. In a real implementation, sensor operational condition and hardware are used to execute work instead of the worst-case energy need defining a task's actual energy usage. Furthermore, task dependencies are not taken into consideration by LSA. Finally, the accuracy of projected available energy is crucial to LSA's efficacy, which is complex and prone to mistakes, as previously stated.

The authors in [49] developed the STAM-STFU protocol, which combines the operations of two scheduling algorithms, the smooth to average method (STAM) and smooth to full utilisation (STFU), to schedule a group of tasks offline with the goal of minimising overall task deadline breaches. Without depending on any energy prediction model, STAM-STFU manages energy uncertainty and deadline limitations. In order to smooth out long-term energy usage, STAM-STFU introduces the notion of virtual tasks. Each unit of actual (physical) work has a virtual counterpart with the same arrival time but a longer or shorter duration and a lower or equivalent energy requirement. Virtual jobs have a longer execution time than their physical equivalents, yet they require the same amount of energy. Virtual task scheduling that adheres to deadline limitations will not interfere with the completion of any actual work. STAM-STFU outperforms non–energy-aware static scheduling methods, according to simulation data. It has also been demonstrated that its performance is comparable to that of LSA, with the added bonus of not necessitating the use of a prediction model.

The multi-version scheduling method [46] aims to complete the most critical and useful periodic activities while keeping all temporal and energy restrictions in mind. It is believed that each task has numerous variants, each with its own set of qualities and rewards. "Easier" variants of a task complete the work faster, use less energy, and give less precise and valued outcomes.

In [50], the authors described earliest deadline with energy guarantee (EDeg), an online scheduling system that is a version of the earliest deadline first method. EDeg ensures energy neutrality by ensuring that enough energy is stored before a task is initiated to cover all subsequent tasks. This protocol implies that the arrival times of future tasks are known. Part of the support is delayed until fuel has been generated by recharging to meet the task deadline. EDeg suspends current processes and begins recharging the battery to a level that permits the activity to be finished when the battery's accumulated energy goes below a particular level. As a result, chores are never completed when there is insufficient energy.

The occurrence utilisation sequence—dependent on the setup of network optimisation by Steck et al. [47] has two main goals: obtain accurate execution resource utilisation of a challenge for a given amount and accurately predict execution over all energy consumption of a group of tasks given a given amount of value. Second, consider which selection of jobs would be more beneficial if you had a short deadline. Under energy constraints, this technique controls assignments by regulating task productivity and team performance, resulting in an energy-neutral system. A directed acyclic graph (DAG) is used to model the interaction between the activities. Additionally, timeframes for task completion, past energy-harvesting data, task requirements, and utility relationships are all supplied in advance. Utilisation is determined for most technologies as a function of job priority and accuracy. A task with a higher priority is accomplished with more efficiency.

The energy-aware DVFS (EA-DVFS) methodology is a continuously adjustable scheduling system that stores and harvests energy for the future [51]. In AS-DVFS [52], adaptive scheduling and the DVFS algorithm were introduced. When possible, AS-DVFS adaptively optimises a node processor's operation frequency and voltage while keeping time and energy constraints in mind. The purpose of AS-DVFS is to reduce system-wide energy consumption by planning and executing tasks as slowly as feasible and distributing the load as equally as possible throughout the processor. It also distinguishes between time and energy constraints, allowing them to be addressed directly. [45] proposes a harvesting-aware DVFS (HA-DVFS) algorithm to increase EA-DVFS and AS-DVFS system performance and energy efficiency. The key aims of HA-DVFS, in particular, are to keep task running speeds as low as feasible while avoiding squandering gathered energy. HA-DVFS schedules jobs and modifies the system's speed and workload to minimise energy overflow based on the predicted energy-harvesting rate in the near future. Distinct time series prediction methodologies have been used to predict the collected energy: regression analysis, moving average, and exponential smoothing. HA-DVFS and AS-DVFS decouple energy and time needs to reduce the complexity of the scheduling algorithm.

Another DVFS-based job scheduling approach is presented in [53]. A linear regression model is used in conjunction with the fundamental DVFS concepts. The protocol's major goals are to maximise system performance given current energy availability, increase energy use efficiency, and improve job accuracy. Because the events created by these types of applications are primarily recurring chores rather than occasional externally driven events, the protocol is judged especially ideal for structural health monitoring applications.

### 2.3.12 Protocols for Network Task Scheduling

At the network level, task allocation entails matching a WSN's sensing resources to relevant tasks (missions) that may arrive on the network dynamically. This is a difficult endeavour since a single node may serve a variety of missions with varying degrees of precision and fit (utility). In relation to the objective (profit) and resources required, operations may differ. They might also enter the network at any moment and for varied amounts of time. Although there have been solutions for WSNs with battery-operated nodes [54], networks with energy collecting nodes have gotten little attention until recently [55]. For these systems, new job allocation concepts are expected that carefully consider the reality that nodes with very little energy left today may have enough sources of energy to carry out new tasks in future.

EN-MASSE [55] is a decentralised technique for sensor-mission deployment in power generation wireless networks that uses energy recovery order features to determine which station should be allocated to which purpose at any given moment. EN-MASSE was designed to recognise and allocate jobs. Each mission joins the network at a different geographic point. During the bidding phase, each node that receives the leader's mission advertisement message determines whether to bid on participation in the mission. This decision is made in accordance with the node's bidding strategy. EN-MASSE classifies activities and evaluates if a node would get energy from the ambient source using an energy prediction model. EN-MASSE can be used with a variety of predictors.

### 2.3.13 Protocols for Harvesting-Aware Communication: MAC and Routing

WSN communication protocols' specifications have switched from energy conservation to pre-emptive optimisation of gathered energy use as a result of harvesting capabilities. The creation of new network topologies is required as a result of this massive change. The goal of this section is to look at WSN MAC and routing solutions. MAC protocols are a group of protocols that are used to communicate over a network.

Nodes that want to send data listen to the channel for a beacon. When a beacon is received, the transmitter attempts to send a packet to the source of the signal. When the beacon duration is short, it takes higher energy for beacon transmission. When the beacon duration is long, more energy is required for beacon transfer. Longer beacon times lead to greater energy saving. The on-demand MAC protocol (ODMAC) protocol's operating mechanism is depicted in Figure 2.7. The sensing and beacon durations of each node are modified on a regular basis based on the current power harvesting rate in ODMAC's dynamic duty cycle mode. In ODMAC,

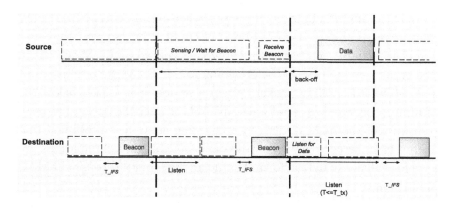

FIGURE 2.7   ODMAC: data packet transmission

harvesting is meant to be irrespective of node behaviours; thus, a sensor could harvest power generated while resting, listening, communicating, and so on. Because it does not recognise and retransmit packets, ODMAC is unsuitable for usage in a lossy environment [56].

Energy adaptive MAC protocol (EA-MAC) [57–59] is an RF energy transfer MAC protocol for energy harvesting-based WSNs (EHWSNs.) The EA-MAC approach is similar to the unslotted carrier sense multiple access (CSMA)/control access (CA) algorithm in IEEE 802.15.4 [60], except the average quantity of harvestable energy determines the sleep length, back-off intervals, and state transitions. The acquired energy level of a node equals the energy required to deliver a packet, and the node switches from sleep to active mode. The packet is then sent using a CSMA/CA method. If the channel is congested, the node chooses between doing the random backoff method and terminating the CSMA/CA algorithm. The number of backoff slots is determined by the current pace of energy collecting. Simulations are used to evaluate analytical models for EA-MAC throughput and fairness [58]. EA-MAC fails to take into account end-to-end latency, lacks a method for optimising network performance and longevity, and suffers from the concealed terminal problem.

The probabilistic polling MAC (PP-MAC) protocol [61] is a polling-based MAC mechanism that employs EHWSN energy properties to optimise the throughput, fairness, and scalability of traditional polling approaches. [62] describes a polling process that is analogous to PP-MAC: The chance of contention is calculated using the current energy-harvesting rate, the number of nodes, and the number of packet collisions. When no sensor replies to the polling packet, the probabilistic polling protocol steadily raises the contention probability. It reduces it if two or more sensor nodes collide. Furthermore, when average energy-harvesting rates increase or drop, the likelihood of conflict increases or decreases. The charge-and-spend harvesting approach is used by PP-MAC, which first gathers enough energy before going to the receive state to listen for and receive the polling packet. When a node's energy falls below the minimum necessary to send a data packet, or after the packet has been transmitted, the node returns to a charging state. It is presumed that energy is gathered solely when the device is charging. Simulations are used to evaluate analytical formulae and analyses of PP-throughput MAC performance.

The multi-tier probabilistic polling (MTPP) technique [59] advances PP-MAC-style probabilistic polling to multi-hop data transfer in EHWSNs

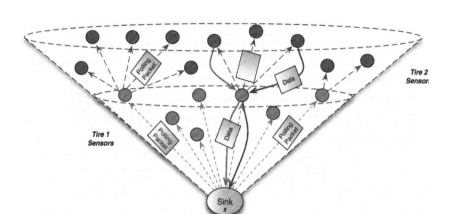

FIGURE 2.8    MTTP multi-tier EHWSN layout

with no battery storage. The polling messages from the sink are provided
to its directly connected neighbours, which then spread them to stations in
succeeding tiers (Figure 2.8). Polling and information packets are broadcast and transferred from layer to layer until they hit their targets. As the
computational complexity rises, the expense of polling packets and packet
collisions increases, leading to higher latencies [63].

### 2.3.14 Routing Protocols

In [64], Pais et al. suggested the hybrid energy storage system (HESS)
routing protocol for hybrid energy storage systems that integrate a super-
capacitor and a rechargeable battery. Their plan was to highlight routes
that use more super-capacitor energy and pass through nodes with significantly greater extracting rates. The main objective of HESS is to bring
down the cost of each end-to-end transfer. When compared to the energy
aware routing (EAR) protocol [65], performance evaluation shows that
HESS increases network residual energy by 10% on average without affecting delivery ratio.

The distributed energy harvesting–aware routing method (DEHAR)
[66] is a routing algorithm for EHWSNs that predicts the fastest route
to the sink depending on sequence number and node energy availability. The penalty should ideally be 0 when the node's energy buffer is fully
charged, and it should trend to infinity when the node's energy is depleted.
When a node's local penalty changes, it broadcasts the change to its near
neighbours.

Energy-harvesting opportunistic routing (EHOR) [67] is a routing system for energy harvesting only. The duty cycle–based adaptive topological (D-APOLLO) knowledge range (KR) algorithm [68] is a harvesting-aware spatial routing system proposed by Noh and Yoon. They intend to increase the utilisation of collected energy while lowering delay by routinely changing the duty cycle and knowledge range of each node. The topographical extent of the data collected by each component is its understanding range. In battery-operated WSNs, the nodes' duty cycle and knowledge range are generally fixed. D-APOLLO, on the other hand, strives to determine the duty cycle and knowledge range that maximises the use of harvested energy based on the projected harvesting power rate, the node's residual energy, and the forecast energy consumption on a periodic basis.

The energy-opportunistic–weighted minimum energy (E-WME) [69] metric determines the shortest path to the sink by taking into account the node's transmission power, memory size, power rate, and required resources for transmitting and receiving messages. The cost of each node is an inversely linear function of the rate of harvesting power, an increasing parameter of residual energy, and a function of transmit and receive energies. In [70], Zeng et al. present two geographic routing algorithms, GREES-L and GREES-M, that take into account energy-harvesting conditions and network quality. Each node must keep track of the position, residual energy, energy-harvesting rate, energy consumption rate, and wireless network quality of its one-hop neighbour nodes. When delivering data to the destination, participating nodes attempt to manage power consumption among their neighbouring nodes by reducing objective functions based on the data they hold. GREES-L and GREES-M only look at possible relay neighbour nodes that offer positive advancement towards the sink, as is typical of greedy geographic forwarding. Simulations show that GREES-L and GREES-M are more energy efficient than the corresponding residual energy–based approaches.

Doost et al. propose an original routing metric based on the charging capacities of sensor nodes in [71]. The metric can be used in conjunction with current routing algorithms in wireless ad hoc and sensor networks. This point is illustrated by work on EHWSNs powered by wireless energy transfer and using the well-known ad hoc routing protocol ad-hoc on demand distance vector [72].

## 2.4 SOURCES OF AMBIENT ENERGY HARVESTING FOR IoT-ENABLED WSNS

The concept of the Internet of Things is widely acknowledged as a vital technical element of the rapidly evolving smart and computer world. IoT components are being developed because of their tremendous sensing and linking capabilities. Embedded devices [73], security monitoring systems [74], control systems [75], transport networks [76], wearable technology [77], power generation tracking [78], climate monitoring systems [79], smoke detectors [80], object recognition [81], vehicle tracking [82], smart agriculture [83, 84], human physical area networking [85], and so on are just some of the applications that IoT can be used for. It is difficult to supply a consistent power source for devices, which may hinder the IoT industry's growth [86].

IoT devices increasingly rely on batteries as their major source of power. The usage of batteries in distant regions may be difficult due to problems such as restricted energy sources, environmental constraints, reduced battery lifetime, and battery maintenance. Energy harvesting is seen to be an effective option for addressing all of these constraints. EH refers to a method of extracting energy from diverse sources. Due to several of energy densities of ambient sources of energy, hybrid energy-harvesting systems (HEHSs) play an essential role in the IoT market. Two energy sources are used in HEHSs to improve system performance and efficiency [87]. The micro-electro-mechanical system (MEMS) production process is the most prevalent type of HEHS in the context of IoT devices. This is referred to as a tiny or micro-scale energy collection system [88].

### 2.4.1 The Motivation of Energy Harvesting in IoT

The Internet of Things timeline is depicted in Figure 2.9. In 1969, the Advanced Research Projects Agency Network (ARPANET), the internet's earliest technological foundation, was built, paving the way for existing internet-controlled devices. One of the first internet-connected devices was a vending machine created by Carnegie Mellon University in 1982. The toaster invented by John Romkey in 1990 [89] was the first internet-controlled device. In the year 2000, LG Electronics Inc. released the first smart refrigerator.

Artificial intelligence (AI), big data, and blockchain-integrated Internet of Things devices have all become increasingly popular since 2017. The majority of enclosed systems are internet connected and easy to use,

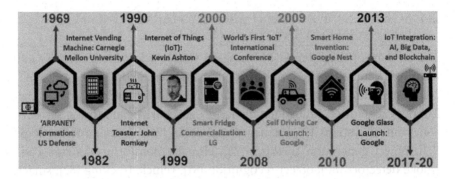

FIGURE 2.9   The Internet of Things timeline (1969–2020)

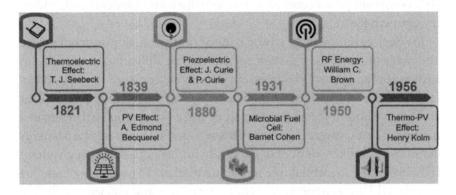

FIGURE 2.10   Chronology of basic energy harvesting (1821–1956)

but some are wireless sensor devices that are small and thus difficult to reach. As a result, these Internet of Things nodes can only be charged by small batteries, which isn't a viable solution because battery installation is expensive and difficult, and it necessitates the use of qualified workers. In these circumstances, developing EH innovations from intermittent energy is a viable method for addressing challenges related to the energisation of IoT-controlled products [90].

## 2.4.2  Principles for Energy Harvesting in IoT Devices

Between 1821 and 1956, the fundamental aspect of energy sources was formulated. Figure 2.10 depicts the fundamental EH history, demonstrating that the EH drift began in 1821, when Thomas Johann Seebeck discovered the thermogenic effect. Alexandre Edmond Becquerel invented the photovoltaic (PV) phenomenon in 1839. (Becquerel provided the photo.) The piezoelectric phenomenon was discovered in 1880 by Jacques and Pierre Curie. In 1931, Barnet Cohen invented the MFC. William C. Brown invented radio frequency in 1950. Henry Kolm discovered the thermo-photovoltaic (thermo-PV) phenomenon in 1956. The major EH movement arose in the mid-1990s, when researchers became more focused on self-sufficient energy approaches for computer parts.

### 2.4.3 Energy-Harvesting Efficiencies

Kymissis and his colleagues created a piezoelectric conversion that could have been stored in a running shoe. It was the first time an environmentally friendly energy harvester was used to power electrical devices [91]. The field of self-powered electrical items has grown to include wireless modules and battery alternatives [92]. Energy-harvesting systems (EHSs) based on a transistor and MEMS have gained popularity [93]. The fundamental goal of EH study was to downsize EH systems to the micro scale and integrate them with flexible electronics [94].

Researchers used dependability issues, size optimisation, feasibility analysis, climatic conditions, and dynamic response to compute the effectiveness of electricity generation [95]. The use of software tools to build an effective solution that decreased the cost of power interruptions was demonstrated. This might be an economically friendly and flexible technology that could be used in the future. Based on various control methodologies, load management systems, economic balance, and so on, a software tool for increasing system performance may be designed.

Because the process is improved, system expenses are reduced [96]. Khan et al. [97] established an HRES-reverse osmosis (HRES-RO) model that used two renewable resources, wind and PV, to enhance EH efficiency and minimise emissions of greenhouse gases. Kasseries et al. [98] proposed a novel hybrid model that combined a wind turbine (WT) with a hydrogen fuel cell (HFC) to boost power generating efficiency. Researchers developed a mathematical model for wind HRES and an isolated PV system in terms of cost efficiency while boosting stability and productivity [99]. Using a register allocation approach, Liu et al. discovered a slew of novel non-volatile processor (NP)–based ambient EHSs. By decreasing critical data overflow, the model and method enhanced system reliability [100].

## 2.4.4 Energy Storage in EH Systems

HRES optimisation strategies were investigated by Bhandari et al. [101] Hydro, wind, and solar power are examples of renewable and alternative energy sources that have gained considerable attention. To make the system stable and cost effective, several optimisation approaches like battery banks or fuel cells can be deployed [101]. The HRES reported by Shivarma et al. utilised fuel cells, wind, and solar energy [87]. The internet of hybrid energy-harvesting things (IoHEHT) was developed by Akan et al. as a self-contained, battery-free system for recharging the batteries of electronic devices [102]. Shi et al. developed a micro-nanosystem using a triboelectric nanogenerator (TENG) and flexible electrical technologies in another work. Because of contemporary TENG and adaptable electrical innovation, the system has become smarter and more efficient. These wearable electronic devices allow for communicating effectively between the body and other multifunctional devices [88]. Perez-Collazo et al. investigated technology that combined offshore wind and wave energy and was tailored to lower industrial costs. This study focused on two topics: wave-to-energy transformation and a substructure [103]. The authors developed a simulation model with an evolutionary method for determining the optimal HES component size. Their strategy included hydrogen storage power stations, wind data warehouses, and fuel cells. The total refined cost of this system was compared to a reasonable battery capacity and hydrogen system. [104] Their method increased efficiency while lowering investment costs. Vosen et al. introduced a methodology for a battery energy storage that is time dependent (HESS).

They compared the neural net control system (NNCS) to the classic cell state-of-charge leadership structure and found that the NNCS was more efficient [105]. A review of the various EH and storage media methods that are employed in IoT devices has been provided, too. They believe that EH systems might be utilised to extend the service life and productivity of IoT devices [106]. The authors in [107] presented a theoretical virtualisation technology for off-loading in WSNs and other sensing devices by employing miniature photovoltaic EH systems. Researchers created an HSPICE prototype system for a micro-power administration system based on a generic hill-climbing technique. They used a 0.35-m complementary metal-oxide-semiconductor (CMOS) to validate the control approach [108]. Previous research [109–112] emphasised the EHS's successes for limited power improvements from a technological standpoint. Table 2.3 displays the results based on the materials utilised in the products, their essential parts, and the merits of micro-scale energy recovery.

TABLE 2.3   Content Perspectives to Power Generation Harvesting Progress

| Ref. | Year | Source | Materials Approaches | Advantages | Obtained Results |
|---|---|---|---|---|---|
| [109] | 2007 | Thermoelectric | ErAs:InGaAs/ (InGaAs) (InAlAs) | Thermodynamically steady superlattice, augmented thermopower coefficient, lower electrical energy loss | $2500 \text{ mW/cm}^2$ (at 3.5 V) |
| [110] | 2007 | Thermo-photovoltaic | p-GaAs/p-Ge/n-Ge cell structure | Solid cell coat with optimised bandgap energy, permit high light source absorbance | $2.5 \text{ W/cm}^2$ (at 3.5 V, 20 μm thickness) |
| [111] | 2008 | Piezoelectric | KNN/Mn/KCT material | Lead-free, high curie temperature, higher density, and piezoelectric coefficient | $10,000 \text{ mW/cm}^3$ |
| [112] | 2008 | Microbial FC | Anode-cathode shallow area distinction | Smooth electron flow, reduced inner resistance | $6.86 \text{ W/m}^2$ (at $2.62 \text{ mA/cm}^2$) |

## 2.4.5 Power Requirement of IoT Sensors and Devices

It is important to understand the overall power concentrations while investigating IoT device implementations. The average output densities are also divided into categories. The normal power concentration of these machines can be categorised using four major existing renewable technologies, as shown in Figure 2.11: ambient light, vibrational/motion, thermal, and radio frequency. In a broad sense, by adequately incorporating natural energy, the difference between current densities and available energy providers can be effortlessly traversed [113–116].

The most power may be obtained by employing a light source. However, because the density of gathered energy may readily fluctuate between 10 mW/cm² and 10 W/cm², conditions can change between indoor and outdoor sites. On a smaller scale, this is equal to vibration and temperature energy sources. Power densities from the two sources of energy, on the other hand, are acquired and employed for commercial processes (100 W/cm² and 1–10 mW/cm², respectively). A commercial macro-EHS generates more than 100 mW of energy, whereas miniaturised EHSs create just 1 mW [92]. Figure 2.12 depicts the amount of energy that might have been created intermittently by the most common sources of ambient energy. The power varies from 0.1 to 100 W at its lowest setting [117]. A temperature generator may generate energy ranging from 10 to more than 100 watts and can be utilised for a variety of purposes. PV-generated EH has grown in popularity in recent years. This energy source produces a large amount of energy, ranging from microwatts to megawatts. This energy source is now being researched, although RF sources have a low yield. This power supply has outputs ranging from 0.1 W to megawatts. Piezoelectric energy produces a large amount of

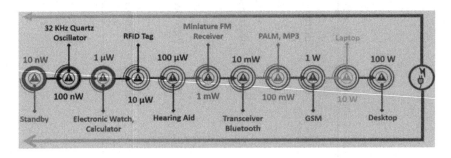

FIGURE 2.11 Different technology sensors and devices with different power demands (10 nW–100 W)

power, ranging from 10 to 100 watts. This type of energy has a great deal of potential for industrialisation.

PV energy is often regarded as the most efficient kind of electricity and is employed in a wide range of IoT applications. Wind energy is employed in a variety of building automation, home automation, and smart city applications. TEGs enhance smart buildings, industrial equipment, and sensors.

### 2.4.6 Energy Harvesting in IoT

In the Internet of Things, sustainable energy includes a range of common forms of energy as well as a few transduction techniques (TTs). Figure 2.13 shows the IoT taxonomy for energy harvesting, which is separated into two classes: ambient based and human body based.

| Power Range | 0.1 µW | 1 µW | 10 µW | 100 µW | 1 mW | 10 mW | 100 mW | 1 W | 10 W | 100 W | <100 W | Industrialization |
|---|---|---|---|---|---|---|---|---|---|---|---|---|
| | | | | Thermoelectric Generator (TEG) | | | | | | | | Widespread Production |
| | | | | Photovoltaic (PV) Energy | | | | | | | | Widespread Production |
| | | Pyroelectric Energy | | | | | | | | | | Research Stage |
| | | Radio Frequency (RF) Waves | | | | | | | | | | Limited Production |
| | | | | Triboelectric Energy | | | | | | | | Research Stage |
| | | | | Piezoelectric Energy | | | | | | | | Widespread Production |

FIGURE 2.12   Power generated occasionally from common ambient energy sources

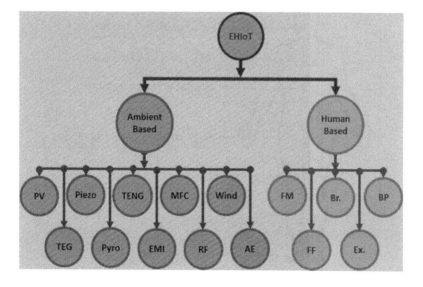

FIGURE 2.13   Taxonomy of energy harvesting in IoT

Finger motion, footfalls, breathing, exhalation, and blood pressure are examples of existing wearable technology based on human-centred energy harvesting in the IoT.

### 2.4.7 Ambient Energy Sources from the Human Body for the IoT

The human body can generate energy in a variety of ways, including hand movement, footsteps, breathing, and blood pressure. In recent years, several devices for monitoring human body parts have been transplanted using wireless body area networks (WBANs) [118, 119]. The amount of energy taken from a human body is seen in Figure 2.14. The available power is compared to the least useable power that can be derived from a body in Table 2.4. This section covers a few energy recovery sources relating to the human body.

Finger motions can be viewed as a reliable type of power for IoT sensors. It has been found that striking the toggle on the Handy Key's Twiddler keyboards requires 130 g finger pressure, which contributes to the emergence of 6.9 mW power. The foundational fall of a heel generates 67b W power on a 5-cm area. Energy may also be retrieved by using breathing. This involves the usage of a mask from outside. Aircraft pilots, astronauts, and other specialists may extract a large quantity of energy from service

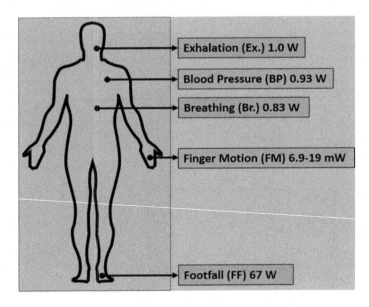

FIGURE 2.14 Different energy-harvesting sources generated from a human body

TABLE 2.4    Description of the Power Generated by Human-Powered Sources

| Human Body-Driven Sources | Available Power | Usable Power |
| --- | --- | --- |
| Finger Motion | 6.9–19 mW | 0.76–2.1 mW |
| Breathing | 0.83 W | 0.42 W |
| Footfall | 67 W | 5.0–8.3 W |
| Exhalation | 1.0 W | 0.40 W |
| Blood Pressure | 0.93 W | 0.37 W |

masks. Breathing helps them develop 0.83 W of power when a band is adjusted across their chest.

## 2.5 CHALLENGES AND RECOMMENDATIONS

Many EH systems have been activated as a result of the decline of conventional energy sources. Because the number of IoT sensors and devices has increased, additional power production is also desirable. The vast majority of ambient power sources are inadequate and incapable of transferring or disseminating collected energy to IoT sensor nodes. As a result, EH efficiency and transduction mechanisms must be enhanced. More dependable and cost-effective transduction technologies for creating vibration EH must be created as the need for micro-scale electricity production devices develops. To ensure that energy is correctly gathered, the accuracy and efficiency of maximum power point tracking must be increased. Some academics are also looking at the auto EH selection process when an energy source is present. It is necessary to build better IoT sensors with a reduced footprint and various inputs that can target tiny devices. IoT nodes will be highly cost effective and viable. Furthermore, because wired energy sources for IoT devices are unavailable, an IoT system has been designed for most battery-powered devices. However, changing the battery is a time-consuming and difficult procedure. As a result, an EH system based on ambient energy is required. The use and efficiency of the systems determine the availability and design of various ambient energy sources. Auto-prediction of user settings lets the consumer feel more at ease and should be looked into further. Simulation tools that are less expensive and more widely available can help researchers and practitioners better understand and apply IoT networks. In light of this, researchers suggest using an open-source simulation platform.

Human body-based EH (HBEH) systems are difficult to implement. Despite its limitations due to adoption, the body area network has grown in popularity among academics. Another constraint that has hampered the widespread adoption of the HBEH method is irregular and non-periodic energy generation. The development of an energy-efficient network protocol is required. As a result, quick inspections, guidelines, and actions are necessary to put in place an effective recycling system. IoT hubs are extremely reliable and long lasting, and they may reduce carbon emissions by mitigating global warming and hazardous battery chemicals.

## 2.6 CONCLUSION

The concept of power generation in WSNs was covered in this chapter, commencing with the design of an energy harvesting–based WSN node and its energy component, as well as protocols for job allocation, MAC and routing, and models for estimating energy availability. Energy harvesting in WSNs is destined to become the leading technology for a diverse range of products that require a network connection for years with breakthroughs in energy recovery techniques and the growth of small- and medium-factor harvesters for a number of energy sources. Energy recovery systems for IoT sensors and devices will require enhanced solutions to fulfil energy demands in the future. In the future, several stockholders are considering energy-harvesting solutions. This chapter has investigated various energy-harvesting sources and strategies that might impact the energy generation process for IoT sensor nodes. Energy recovery levels, collective architecture, harvested energy level evaluation, and harvester capabilities of all classes and subclasses have all been researched. Each energy source should have its own set of harnessing abilities, and researchers should discover how to maximise the huge potential of varied energy sources. In addition, a detailed evaluation of many energy-harvesting models was performed in order to evaluate the probable energy cycles for handling issues with an intermittent power supply for powering IoT nodes. Finally, many unresolved difficulties with non-exclusive energy harvesters used in diverse contexts were addressed.

## REFERENCES

[1] E. O. Torres and G. A. Rincón-Mora, "Energy-harvesting chips and the quest for everlasting life," *IEEE Georgia Tech Analog and Power IC Design Lab*, 2005.

[2] S. Mikami, M. Tetsuro, Y. Masahiko and O. Hiroko, "A wireless-interface SoC powered by energy harvesting for short-range data communication," *IEEE 0-7803-9162-4/05*, 2005.

[3] T. Sterken, P. Fiorini, K. Baert, R. Puers and G. Borghs, "An electret-based electrostatic μ-generator," *IEEE 0-7803-7731-1/03*, 2003.

[4] S. Roundy, P. K. Wright and J. Rabaey, *Energy Scavenging for Wireless Sensor Networks with Special Focus on Vibrations*, New York, 2004.

[5] R. Amirtharajah and A. P. Chandrakasan, "Self-powered signal processing using vibration based power generation," *IEEE Journal of Solid-State Circuits*, vol. 33, no. 5, pp. 687–695, 1998.

[6] H. A. Sodano, D. J. Inman and G. Park, "A review of power harvesting from vibration using piezoelectric materials," *The Shock and Vibration Digest*, vol. 36, no. 3, pp. 197–205, 2004.

[7] S. Roundy, P. K. Wright and J. Rabaey, *Energy Scavenging for Wireless Sensor Networks with Special Focus on Vibrations*, New York, 2004.

[8] D. A. Skoog, J. F. Holler and S. R. Crouch, *Principles of Instrumental Analysis*, 6th edition, Florence, KY, 2006.

[9] S. Meninger, J. O. Mur-Miranda, R. Amirtharajah, A. P. Chandrakasan and J. H. Lang, "Vibration-to-electric energy conversion," *IEEE Transactions on Very Large Scale Integration (VLSI) Systems*, vol. 9, no. 1, pp. 64–76, 2001.

[10] M. Marzencki, "Vibration energy scavenging. European Commission research project," *VIBES (IST-1-507911) of the 6th STREP Framework Program*, 2005.

[11] F. J. DiSalvo, "Thermoelectric cooling and power generation," *Science*, vol. 285, pp. 703–706, 1999.

[12] D. M. Rowe, "Thermoelectrics, an environmentally-friendly source of electrical power," *Renewable Energy*, 16, pp. 1251–1256, 1999.

[13] Pacific Northwest National Laboratory (PNNL), "Available technologies patent pending," *Battelle Number(s)*, pp. 12398-E–13664-B. http://availabletechnologies. pnl.gov/technology.asp?id=85, 2007.

[14] D. Pescovitz, "The power of small technology," *Smalltimes*, vol. 2, no. 1, 2002.

[15] J. Thomas, J. W. Clark and W. W. Clark, "Harvesting energy from piezoelectric material," *IEEE CS. University of Pittsburgh*, pp. 1536–1268, 2005.

[16] S. B. Lang, "Pyroelectricity: From ancient curiosity to modern imaging tool. Changes in the net dipole moment of certain materials form the basis for a broad range of IR detectors," 2005. www.physicstoday.org/vol-58/iss-8/p31. html. Accessed 5 October 2009.

[17] R. Raffaelle, J. Underwood, D. Scheiman, J. Cowen, P. Jenkins, A. F. Hepp, J. Harris and D. M. Wilt, "Integrated solar power systems," in *28th IEEE Photovoltaic Specialists Conference*, pp. 1370–1373, Anchorage, AK, 2000.

[18] W. Sun, N. P. Kherani, K. D. Hirschman, L. L. Gadeken and P. M. Fauchet, "A three-dimensional porous silicon p-n diode for betavoltaics and photovoltaics," *Advanced Materials*, vol. 17, pp. 1230–1233, 2005.

[19] A. L. Rogers, J. F. Manwell and S. P. E. Wright, "Wind turbine acoustic noise," *Renewable Energy Research Laboratory*, Department of Mechanical and Industrial Engineering, University of Massachusetts at Amherst, 2002.

[20] S. Horowitz, A. Kasyap, F. Liu, D. Johnson, T. Nishida, K. Ngo, M. Sheplak and L. Cattafesta, "Technology development for self-powered sensors," in *Proceedings of 1st Flow Control Conference*, St Louis, 2002.

[21] A. Kasyap, J. S. Lim, D. Johnson, S. Horowitz, T. Nishida, K. Ngo, M. Sheplak and L. Cattafesta, "Energy reclamation from a vibrating piezoceramic composite beam," in *Proceedings of 9th International Conference on Sound and Vibration*, Orlando, 2002.

[22] T. Starner and J. A. Paradiso, "Human-generated power for mobile electronics," in C. Piguet (Ed.), *Low-Power Electronics Design, Chapter*, vol. 45, pp. 1–35, New York, NY, 2004.

[23] R. Adler, P. Desmares and J. Spracklen, "Ultrasonic remote control for home receivers," *IEEE Transaction Consumer Electronics*, vol. 28, no. 1, pp. 123–128, 1982.

[24] J. A. Paradiso and M. Feldmeier. "A compact, wireless, self-powered pushbutton controller," in G. D. Abowd, B. Brumitt and S. Shafer (Eds.), *Ubicomp 2001: Ubiquitous Computing. UbiComp 2001. Lecture Notes in Computer Science*, vol. 2201, pp. 299–304, 2001, Berlin, Heidelberg. https://doi.org/10.1007/3-540-45427-6_25.

[25] N. S. Shenck and J. A. Paradiso, "Energy scavenging with shoe-mounted piezoelectric," *IEEE Micro*, vol. 21, pp. 30–41, 2001.

[26] S. Sudevalayam and P. Kulkarni, "Energy harvesting sensor nodes: Survey and implications," *IEEE Communications Surveys Tutorials*, vol. 13, no. 3, pp. 443–461, Third quarter 2011.

[27] T. Zhu, Z. Zhong, Y. Gu, T. He and Z.-L. Zhang, "Leakage-aware energy synchronization for wireless sensor networks," in *Proceedings of ACM MobiSys*, pp. 319–332, New York, NY, 2009.

[28] F. Simjee and P. H. Chou, "Everlast: Long-life, supercapacitor-operated wireless sensor node," in *Proceedings of ISLPED 2006*, pp. 197–202, Tagernsee, Germany, 4–6 October 2006.

[29] J. C. Park, D. H. Bang and J. Y. Park, "Micro-fabricated electromagnetic power generator to scavenge low ambient vibration," *IEEE Transactions on Magnetics*, vol. 46, no. 6, pp. 1937–1942, June 2010.

[30] P. Corke, T. Wark, P. Valencia and P. Sikka, "Long-duration solar-powered wireless sensor networks," in *Proceedings of EmNets 2007*, pp. 33–37, Cork, Ireland, 25–26 June 2007.

[31] G. V. Merrett, A. S. Weddell, A. P. Lewis, N. R. Harris, B. M. Al-Hashimi and N. M. White, "An empirical energy model for supercapacitor powered wireless sensor nodes," in *Proceedings of IEEE ICCCN*, pp. 1–6, US Virgin Islands, 3–7 August 2008.

[32] A. Kansal, J. Hsu, S. Zahedi and M. B. Srivastava, "Power management in energy harvesting sensor networks," *ACM Transactions in Embedded Computing Systems*, vol. 6, no. 4, Article 32, September 2007.

[33] R. Rao, S. Vrudhula and D. N. Rakhmatov, "Battery modelling for energy aware system design," *Computer Magazine*, vol. 36, no. 12, pp. 77–87, December 2003.

[34] M. Pedram, "Design considerations for battery-powered electronics," in *Proceedings of Design Automation Conference*, pp. 861–866, New Orleans, LA, 21–25 June 1999.

[35] C. F. Chiasserini and R. R. Rao, "Energy efficient battery management," *IEEE Journal on Selected Areas in Communications*, vol. 19, no. 7, pp. 1235–245, July 2001.

[36] H. J. Bergveld, W. S. Kruijt and P. H. L. Notten, "Electronic-network modelling of rechargeable NiCd cells and its application to the design of battery management systems," *Journal of Power Sources*, vol. 77, no. 2, pp. 143–158, March 1999.

[37] L. Benini, G. Castelli, A. Macii, E. Macii, M. Poncino and R. Scarsi, "A discretetime battery model for high-level power estimation," in *Proceedings of DATE 2000*, pp. 35–41, Paris, France, 27–30 March 2000.

[38] D. N. Rakhmatov and S. B. K. Vrudhula, "An analytical high-level battery model for use in energy management of portable electronic systems," in *Proceedings of IEEE/ACM ICCAD 2001*, pp. 488–493, San Jose, CA, 4–8 November 2001.

[39] D. R. Cox, "Prediction by exponentially weighted moving averages and related methods," *Journal of the Royal Statistical Society. Series B (Methodological)*, vol. 23, no. 2, pp. 414–422, 1961.

[40] J. R. Piorno, C. Bergonzini, D. Atienza and T. S. Rosing, "Prediction and management in energy harvested wireless sensor nodes," in *Proceedings of Wireless VITAE*, pp. 6–10, 17–20 May 2009.

[41] C. Bergonzini, D. Brunelli and L. Benini, "Comparison of energy intake prediction algorithms for systems powered by photovoltaic harvesters," *Microelectronics Journal*, vol. 41, no. 11, pp. 766–777, November 2010.

[42] C. Moser, L. Thiele, D. Brunelli and L. Benini, "Adaptive power management in energy harvesting systems," in *Design, Automation & Test in Europe Conference & Exhibition*, France, pp. 1–6, 2007. https://doi.org/10.1109/DATE.2007.364689.

[43] A. Cammarano, C. Petrioli and D. Spenza, "Pro-energy: A novel energy prediction model for solar and wind energy harvesting *Wireless Sensor Networks*," in *2012 IEEE 9th International Conference on Mobile Ad-Hoc and Sensor Systems (MASS 2012)*, NJ, USA, pp. 8–11, October 2012.

[44] N. Sharma, J. Gummeson, D. Irwin and P. Shenoy, "Cloudy computing: Leveraging weather forecasts in energy harvesting sensor systems," in *Proceedings of SECON 2010*, pp. 1–9, Boston, MA, 21–25 June 2010.

[45] S. Liu, J. Lu, Q. Wu and Q. Qiu, "Harvesting-aware power management for real-time systems with renewable energy," *IEEE Transactions on Very Large Scale Integration (VLSI) Systems*, vol. 20, no. 8, pp. 1473–1486, August 2012. https://doi.org/10.1109/TVLSI.2011.2159820.

[46] C. Rusu, R. Melhem and D. Moss_e, "Multi-version scheduling in rechargeable energy-aware real-time systems," *Journal of Embedded Computing*, vol. 1, no. 2, pp. 271–283, April 2005.

[47] J. B. Steck and T. S. Rosing, "Adapting task utility in externally triggered energy harvesting wireless sensing systems," in *2009 Sixth International Conference on Networked Sensing Systems (INSS)*, Pennsylvania, USA, pp. 1–8, June 2009. https://doi.org/10.1109/INSS.2009.5409959.

[48] C. Moser, J.-J. Chen and L. Thiele, "Dynamic power management in environmentally powered systems," in *2010 15th Asia and South Pacific Design Automation Conference (ASP-DAC)*, Taiwan, pp. 81–88, January 2010. https://doi.org/10.1109/ASPDAC.2010.5419916.

[49] D. Audet, L. C. de Oliveira, N. MacMillan, D. Marinakis and K. Wu, "Scheduling recurring tasks in energy harvesting sensors," in *2011 IEEE Conference on Computer Communications Workshops (INFOCOM WKSHPS)*, Shanghai, China, pp. 277–282, April 2011. https://doi.org/10.1109/INFCOMW.2011.5928823.

[50] H. E. L. Ghor, M. Chetto and R. H. Chehade, "A real-time scheduling framework for embedded systems with environmental energy harvesting," *Computers and Electrical Engineering Journal*, vol. 37, no. 4, pp. 498–510, July 2011.

[51] S. Liu, Q. Qiu and Q. Wu, "Energy aware dynamic voltage and frequency selection for real-time systems with energy harvesting," in *2008 Design, Automation and Test in Europe*, Munich, Germany, pp. 236–241, March 2008. https://doi.org/10.1109/DATE.2008.4484692.

[52] S. Liu, Q. Wu and Q. Qiu, "An adaptive scheduling and voltage/frequency selection algorithm for real-time energy harvesting systems," in *2009 46th ACM/IEEE Design Automation Conference*, CA, USA, pp. 782–787, July 2009.

[53] A. Ravinagarajan, D. Dondi and T. S. Rosing, "DVFS based task scheduling in a harvesting WSN for structural health monitoring," in *Proceedings of DATE*, pp. 1518–1523, Leuven, Belgium, 8–12 March 2010.

[54] M. P. Johnson, H. Rowaihy, D. Pizzocaro, A. Bar-Noy, S. Chalmers, T. La Porta and A. Preece, "Sensor-mission assignment in constrained environments," *IEEE Transactions on Parallel and Distributed Systems*, vol. 21, no. 11, pp. 1692–1705, November 2010.

[55] T. La Porta, C. Petrioli and D. Spenza, "Sensor-mission assignment in wireless sensor networks with energy harvesting," in *Proceedings of IEEE SECON 2011*, pp. 413–421, Salt Lake City, UT, 27–30 June 2011.

[56] X. Fafoutis and N. Dragoni, "ODMAC: An on-demand MAC protocol for energy harvesting wireless sensor networks," in *Proceedings of ACM PE-WASUN 2011*, FL, U.S.A, pp. 49–56, 3–4 November 2011.

[57] J. Kim and J.-W. Lee, "Energy adaptive MAC protocol for wireless sensor networks with RF energy transfer," in *2011 Third International Conference on Ubiquitous and Future Networks (ICUFN)*, Dalian, China, pp. 89–94, June 2011. https://doi.org/10.1109/ICUFN.2011.5949141.

[58] J. Kim and J.-W. Lee, "Performance analysis of the energy adaptive MAC protocol for wireless sensor networks with RF energy transfer," in *ICTC*, Seoul, Korea, pp. 14, 19, September 2011. https://doi.org/10.1109/ICTC.2011.6082542.

[59] C. Fujii and W. K. G. Seah, "Multi-tier probabilistic polling for wireless sensor networks powered by energy harvesting," in *2011 Seventh International Conference on Intelligent Sensors, Sensor Networks and Information Processing*, SA, Australia, pp. 383–388, December 2011. https://doi.org/10.1109/ISSNIP.2011.6146627.

[60] Z. A. Eu, H.-P. Tan and W. K. G. Seah, "Design and performance analysis of MAC schemes for wireless sensor networks powered by ambient energy harvesting," *Ad Hoc Network*, vol. 9, no. 3, pp. 300–323, May 2011.

[61] IEEE 802.15.4–2006 Standard, "Wireless medium access control (MAC) and physical layer (PHY) specifications for low-rate wireless personal area networks (WPANs)," 2006.

[62] Z. A. Eu, W. K. G. Seah and H.-P. Tan, "A study of MAC schemes for wireless sensor networks powered by ambient energy harvesting," in *Proceedings of WICON 2008*, pp. 78:1–78:9, Maui, Hawaii, 2008.

[63] F. Iannello, O. Simeone and U. Spagnolini, "Medium access control protocols for wireless sensor networks with energy harvesting," *IEEE Transactions on Communications*, vol. 60, no. 5, pp. 1381–1389, 2012. https://doi.org/10.1109/TCOMM.2012.030712.110089.

[64] N. Pais, B. K. Cetin, N. Pratas, F. J. Velez, N. R. Prasad and R. Prasad, "Cost-benefit aware routing protocol for wireless sensor networks with hybrid energy storage system," *Journal of Green Engineering*, vol. 1, 2, pp. 189–208, January 2011.

[65] P. K. K. Loh, S. H. Long and Y. Pan, "An efficient and reliable routing protocol for wireless sensor networks," in *Sixth IEEE International Symposium on a World of Wireless Mobile and Multimedia Networks*, Italy, pp. 512–516, June 2005. https://doi.org/10.1109/WOWMOM.2005.25.

[66] M. K. Jakobsen, J. Madsen and M. R. Hansen, "DEHAR: A distributed energy harvesting aware routing algorithm for ad-hoc multi-hop wireless sensor networks," in *2010 IEEE International Symposium on "A World of Wireless, Mobile and Multimedia Networks" (WoWMoM)*, Montreal, QC, Canada, pp. 1–9, June 2010. https://doi.org/10.1109/WOWMOM.2010.5534899.

[67] Z. A. Eu, H.-P. Tan and W. K. G. Seah, "Opportunistic routing in wireless sensor networks powered by ambient energy harvesting," *Computer Networks*, vol. 54, no. 17, pp. 2943–2966, December 2010.

[68] D. Noh and I. Yoon, "Low-latency geographic routing for asynchronous energy harvesting," *WSNs. Journal of Networks*, vol. 3, no. 1, pp. 78–85, 2008.

[69] L. Lin, N. B. Shro and R. Srikant, "Asymptotically optimal energy-aware routing for multihop wireless networks with renewable energy sources," *IEEE/ACM Transactions on Networking*, vol. 15, no. 5, pp. 1021–1034, October 2007.

[70] K. Zeng, K. Ren, W. Lou and P. J. Moran, "Energy-aware geographic routing in lossy wireless sensor networks with environmental energy supply," *ACM Wireless Networks (WINET)*, vol. 15, no. 1, pp. 39–51, January 2009.

[71] R. Doost, K.R. Chowdhury and M. Di Felice, "Routing and link layer protocol design for sensor networks with wireless energy transfer," in *2010 IEEE Global Telecommunications Conference GLOBECOM 2010*, FL, U.S.A, pp. 1–5, December 2010. https://doi.org/10.1109/GLOCOM.2010.5683334.

[72] I. D. Chakares and E. M. Belding-Royer, "AODV routing protocol implementation design," in *Proceedings of WWAN 2004*, pp. 698–703, Tokyo, Japan, 23–24 March 2004.

[73] A. Pötsch, A. Berger and A. Springer, "Efficient analysis of power consumption behaviour of embedded wireless IoT systems," in *Proceedings of the 2017 IEEE International Instrumentation and Measurement Technology Conference*, pp. 1–6, Turin, Italy, 22–25 May 2017.

[74] S. Sruthy and S. N. George, "WiFi enabled home security surveillance system using Raspberry Pi and IoT module," in Proceedings of the 2017 IEEE International Conference on Signal Processing, Informatics, Communication and Energy Systems (SPICES), pp. 1–6, Kollam, India, 8–10 August 2017.

[75] M. Bardwell, J. Wong, S. Zhang and P. Musilek, "Design considerations for IoT-based PV charge controllers," in *Proceedings of the 2018 IEEE World Congress on Services (SERVICES)*, pp. 59–60, San Francisco, CA, 2–7 July 2018.

[76] A. J. V. Neto, Z. Zhao, J. J. P. C. Rodrigues, H. B. Camboim and T. Braun, "Fog-based crime-assistance in smart IoT transportation system," *IEEE Access 2018*, vol. 6, pp. 11101–11111, 2018.

[77] T. Nguyen Gia, V. K. Sarker, I. Tcarenko, A. M. Rahmani, T. Westerlund, P. Liljeberg and H. Tenhunen, "Energy efficient wearable sensor node for IoT-based fall detection systems. Microprocess," *Microsystem*, vol. 56, pp. 34–46, 2018.

[78] V. Mani and Abhilasha Gunasekhar, "Iot based smart energy tracking system," *International Journal of Applied Engineering Research*, vol. 12, pp. 5455–5462, 2017.

[79] G. Mois, S. Folea and T. Sanislav, "Analysis of three IoT-based wireless sensors for environmental monitoring," *IEEE Transactions on Instrumentation and Measurement*, vol. 66, pp. 2056–2064, 2017.

[80] K. Muhammad, S. Khan, M. Elhoseny, S. Hassan Ahmed and S. Wook Baik, "Efficient fire detection for uncertain surveillance environment," *IEEE Transactions on Industrial Informatics*, vol. 15, pp. 3113–3122, 2019.

[81] S. Anandhi, R. Anitha and V. Sureshkumar, "IoT enabled RFID authentication and secure object tracking system for smart logistics," *Wireless Personal Communications*, vol. 104, pp. 543–560, 2019.

[82] S. Dhingra, R. B. Madda, R. Patan, P. Jiao, K. Barri and A. H. Alavi, "Internet of Things-based fog and cloud computing technology for smart traffic monitoring," *Internet Things*, p. 100175, 2020.

[83] A. Tzounis, N. Katsoulas, T. Bartzanas and C. Kittas, "Internet of Things in agriculture, recent advances and future challenges," *Biosystems Engineering*, vol. 164, pp. 31–48, 2017.

[84] M. A. Zamora-Izquierdo, J. Santa, J. A. Martínez, V. Martínez and A. F. Skarmeta, "Smart farming IoT platform based on edge and cloud computing," *Biosystems Engineering*, vol. 177, pp. 4–17, 2019.

[85] G. Elhayatmy, N. Dey and A. S. Ashour, "Internet of Things based wireless body area network in healthcare," in *Internet of Things and Big Data Analytics toward Next-Generation Intelligence*, pp. 3–20, Berlin, Germany, 2018.

[86] M. Babar, A. Rahman, F. Arif and G. Jeon, "Energy-harvesting based on Internet of Things and big data analytics for smart health monitoring," *Sustainable Computing: Informatics and Systems*, vol. 20, pp. 155–164, 2018.

[87] K. Shivarama Krishna and K. Sathish Kumar, "A review on hybrid renewable energy systems," *Renewable and Sustainable Energy Reviews*, vol. 52, pp. 907–916, 2015.

[88] Q. Shi, T. He and C. Lee, "More than energy harvesting—Combining tri-boelectric nanogenerator and flexible electronics technology for enabling novel micro-/nano-systems," Nano Energy, vol. 57, pp. 851–871, 2019.

[89] S. Lata, S. Mehfuz and S. Urooj, "Secure and reliable WSN for Internet of Things: Challenges and enabling technologies," in *IEEE Access*, vol. 9, pp. 161103–161128. doi:10.1109/ACCESS.2021.3131367, 2021.

[90] M. Babar, A. Rahman, F. Arif and G. Jeon, "Energy-harvesting based on Internet of Things and big data analytics for smart health monitoring," *Sustainable Computing: Informatics and Systems*, vol. 20, pp. 155–164, 2018.

[91] J. Kymissis, C. Kendall, J. Paradiso and N. Gershenfeld, "Parasitic power harvesting in shoes," in *Proceedings of the Digest of Papers. Second International Symposium on Wearable Computers* (Cat. No. 98EX215), pp. 132–139, Pittsburgh, PA, 19–20, October 1998.

[92] S. Roundy and P. K. Wright, "A piezoelectric vibration based generator for wireless electronics," *Smart Materials and Structures*, vol. 13, p. 1131, 2004.

[93] S. S. Tan, C.-Y. Liu, L.-K. Yeh, Y.-H. Chiu, M. S.-C. Lu and K. Y. J. Hsu, "An integrated low-noise sensing circuit with efficient bias stabilization for CMOS MEMS capacitive accelerometers," *IEEE Transactions on Circuits and Systems. I Regul. Pap.*, vol. 58, pp. 2661–2672, 2011.

[94] J. Iannacci, "Microsystem based energy harvesting (EH-MEMS): Powering pervasivity of the Internet of Things (IoT)—A review with focus on mechanical vibrations," *Journal of King Saud University Science*, vol. 31, pp. 66–74, 2019.

[95] V. Khare, S. Nema and P. Baredar, "Solar-wind hybrid renewable energy system: A review," *Renewable and Sustainable Energy Reviews*, vol. 58, pp. 23–33, 2019.

[96] S. Sinha and S. S. Chandel, "Review of software tools for hybrid renewable energy systems," *Renewable and Sustainable Energy Reviews*, vol. 32, pp. 192–205, 2014.

[97] M. A. M. Khan, S. Rehman and F. A. Al-Sulaiman, "A hybrid renewable energy system as a potential energy source for water desalination using reverse osmosis: A review," *Renewable and Sustainable Energy Reviews*, vol. 97, pp. 456–477, 2018.

[98] E. Kasseris, Z. Samaras and D. Zafeiris, "Optimization of a wind-power fuel-cell hybrid system in an autonomous electrical network environment," *Renewable Energy*, vol. 32, pp. 57–79, 2007.

[99] J. L. Bernal-Agustín and R. Dufo-López, "Simulation and optimization of stand-alone hybrid renewable energy systems," *Renewable and Sustainable Energy Reviews*, vol. 13, pp. 2111–2118, 2009.

[100] Y. Liu, Z. Li, H. Li, Y. Wang, X. Li, K. Ma, S. Li, M. F. Chang, S. John, Y. Xie, et al., "Ambient energy harvesting non volatile processors: From circuit to system," in *Proceedings of the 52nd Annual Design Automation Conference*, pp. 1–6, San Francisco, CA, 8–12 June 2015.

[101] B. Bhandari, K. T. Lee, G. Y. Lee, Y. M. Cho and S. H. Ahn, "Optimization of hybrid renewable energy power systems: A review," *International Journal of Precision Engineering and Manufacturing—Green Technology*, vol. 2, pp. 99–112, 2015.

[102] O. B. Akan, O. Cetinkaya, C. Koca and M. Ozger, "Internet of hybrid energy harvesting things," *IEEE Internet Things Journal*, vol. 5, pp. 736–746, 2018.

[103] C. Pérez-Collazo, D. Greaves and G. Iglesias, "A review of combined wave and offshore wind energy," *Renewable and Sustainable Energy Reviews*, vol. 42, pp. 141–153, 2015.

[104] A. J. del Real, A. Arce and C. Bordons, "Optimization strategy for element sizing in hybrid power systems," *Journal of Power Sources*, vol. 193, pp. 315–321, 2009.

[105] S. R. Vosen and J. O. Keller, "Hybrid energy storage systems for stand-alone electric power systems: Optimization of system performance and cost through control strategies," *International Journal of Hydrogen Energy*, vol. 24, pp. 1139–1156, 1999.

[106] N. Garg and R. Garg, "Energy harvesting in IoT devices: A survey," in *Proceedings of the 2017 International Conference on Intelligent Sustainable Systems (ICISS)*, pp. 127–131, Palladam, India, 7–8 December 2017.

[107] C. Kulatunga, K. Bhargava, D. Vimalajeewa and S. Ivanov, "Cooperative in-network computation in energy harvesting device clouds," *Sustainable Computing: Informatics and Systems*, vol. 16, pp. 106–116, 2017.

[108] H. Shao, C. Y. Tsui and W. H. Ki, "A micro power management system and maximum output power control for solar energy harvesting applications," in *Proceedings of the 2007 International Symposium on Low Power Electronics and Design (ISLPED'07)*, pp. 298–303, Portland, OR, 27–29 August 2007.

[109] G. Zeng, J. H. Bahk, J.-E. Bowers, J. M. O. Zide, A. C. Gossard, Z. Bian, R. Singh, A. Shakouri, W. Kim, S. L. Singer and A. Majumdar, " ErAs: (InGaAs)1-x (InAlAs)x alloy power generator modules," *Applied Physics Letters*, vol. 91, p. 26, 2007.

[110] V. P. Khvostikov, O. A. Khvostikova, P. Y. Gazaryan, S. V. Sorokina, N. S. Potapovich, A. V. Malevskaya, N. A. Kaluzhniy, M. Z. Shvarts and V. M. Andreev, "Photovoltaic cells based on GaSb and Ge for solar and thermophotovoltaic applications," *Journal of Solar Energy Engineering*, vol. 129, pp. 291–297, 2007.

[111] D. Lin, M. S. Guo, K. H. Lam, K. W. Kwok and H. L. W. Chan, "Lead-free piezoelectric ceramic (K0.5Na0.5) NbO$_3$ with MnO$_2$ and K5.4Cu1.3Ta10O29 doping for piezoelectric transformer application," *Smart Materials and Structures*, vol. 17, p. 35002, 2008.

[112] Y. Fan, E. Sharbrough and H. Liu, "Quantification of the internal resistance distribution of microbial fuel cells," *Environmental Science & Technology*, vol. 42, pp. 8101–8107, 2008.

[113] M. L. Pykälä, K. Sipilä, U. M. Mroueh, M. Wahlström, H. Huovila, T. Tynell, T. Tervo and J. Susterel, "Energy harvesting roadmap for societal applications," in *VTT Research Report VTT*, Espoo, Finland, 2012.

[114] M. Hayes, "Synergies between energy harvesting and power electronics slides," https://resourcecenter.ieee-pels.org/webinars/PELSWEB01020 2016.html. Accessed 23 March 2020.

[115] B. Franciscatto, "Design and implementation of a new low-power consumption DSRC transponder," Ph.D. Thesis, Université de Grenoble, Saint-Martin-d'Hères, France, 2014.

[116] M. Shirvanimoghaddam, K. Shirvanimoghaddam, M. M. Abolhasani, M. Farhangi, V. Zahiri Barsari, H. Liu, M. Dohler and M. Naebe, "Towards a green and self-powered Internet of Things using piezoelectric energy harvesting," *IEEE Access 2019*, vol. 7, pp. 94533–94556, 2019.

[117] F. Ünlü, L. Wawrla and A. Dìaz, "Energy harvesting technologies for Iot edge devices; 4E," in *International Energy Agency*, Paris, France, 2018.

[118] B. Latré, B. Braem, I. Moerman, C. Blondia and P. Demeester, "A survey on wireless body area networks," *Wireless Network*, vol. 17, pp. 1–18, 2011. https://doi.org/10.1007/s11276-010-0252-4.

[119] T. Starner, "Human-powered wearable computing," *IBM System of Journal*, vol. 35, pp. 618–629, 1996.

# CHAPTER 3

# Analytical Modelling of HVAC-IoT Systems with the Aid of UVGI and Solar Energy Harvesting

Shafeeq Ahmad, Md. Toufique Alam, Mohammad Bilal, Osama Khan, Mohd Zaheen Khan

## CONTENTS

### *Nomenclature*

| | |
|---|---|
| **AC** | Air conditioning |
| **AHU** | Air handling unit |
| **EB** | Electricity bill |
| **ECBC** | Energy conservation building code |
| **EC** | Electric current |
| **EEM** | Energy efficiency measure |
| **FCU** | Fan coil unit |
| **HP** | Horsepower |
| **HVAC** | Heating, ventilation and air conditioning |

DOI: 10.1201/9781003218760-3

| **IoT** | Internet of Things |
| **UVGI** | Ultraviolet germicidal irradiation |
| **USGPM** | US gallons per minute |
| **VFD** | Variable frequency drive |
| **VRF** | Variable refrigerant flow |

## 3.1 INTRODUCTION

The current eruption of destruction in various countries instigated by the coronavirus (COVID-19) raises grave public concern. Various individuals of different ethnicities across the world have been infected with COVID-19 and furthermore could suffer from extreme respiratory illness. More than 250 nations, zones, or territories have recorded considerable numbers of COVID-19 infections, demonstrating the brutality of the global epidemic [1–2]. The death rate of patients infected with this disease due to the spread of the virus is still rising at an exponential rate and needs to be controlled. Among the most effective and essential techniques to control the spread of this contagious virus in the pandemic are proper medicinal conduct, proper lockdown, work from home, and so on, while social isolation still remains the most effective solution. To prevent the virus from reaching healthy persons, adequate efforts must be applied to counteract it [3–5].

Yet when people return to their normal lifestyles, professions, and production, there will be more people accessing public services (e.g., stores, grocers, workplaces, healthcare facilities, diners, and cinemas), raising the danger of a new epidemic of disease cases [6–9]. Furthermore, recently, the spread of poor construction conditions has become another threat to current resumption of services, as the employment of sanitisers in construction has become excessive. Proper monitoring and modifications within indoor climatic conditions is crucial so as to diminish the hazard of contamination in these defined spaces [10–12].

Heating, ventilation, and air-conditioning (HVAC) systems, properly integrated with each other, provide healthy, safe, and thermally comfortable indoor surroundings. Henceforth, the coherent usage of HVAC regulation is greatly advantageous for the eco-friendly controller to diminish contamination hazards and expand human welfare in extreme conditions. Model-based predictive monitoring (MBPM) integrated with UVGI technology is a technique of HVAC control that offers the largest potential for power savings, with 99% virus elimination efficiency. Despite the substantial amount of research, only a few reports take into consideration MBPM applications for existing buildings under COVID-19 conditions, making this a novel technique in the HVAC area. So far, considering previous literature reviews, this is an original effort towards a saleable MBPM-COVID-19 HVAC possible resolution [13–16].

## 3.2 MATERIALS AND METHODS

So far, viral agents are generally transmitted through coming into contact with the person infected, drop spread or respiratory spread. Airborne transmission through droplet carriers of the virus is the most significant transmission method. Also, these virus agents' nuclei and particulate matter have the potential to cover substantial distances, since these amalgamated particles are carried by airflow or buoyancy effects [17–19]. Eventual exposure to viruses carrying minute particles within enclosed zones would result in a substantially large number of infections among people. The guidelines established by the WHO clearly show proper ventilation of internal zones in combination to regularly following essential special hygiene conditions. Therefore, proper ventilation control and monitoring must be provided so as to fulfil all necessary requirements of successfully lowering the contamination danger, predominantly in closed spaces. Furthermore, an immediate measure regarding continuous improvements in the national standards of HVAC should be updated so as to ensure adequate virus-free ventilation facilities while taking into consideration the cost effectiveness and efficiency of the system. This research intends to describe the strategy and authentication of two types of equipment for this combined arrangement, solar-powered sensors and Internet of Things (IoT) settings. Outcomes generated from response surface methodology (RSM) in an actual setup of an enclosed area under typical conditions can establish a reserve money fund in the energy bill while preserving actual comfort levels and simultaneously eliminating viruses [20].

The following conditions were assumed for calculations for the HVAC system;

1 Fresh air: 7.5 cfm/per person per ASHRAE 170 and 62.1

2 Window glazing: Single/double-pane glazed glass

3 Lighting load: 1.1 W/sq. ft

4 Occupancy: 20–25 people per seating plan

5 Equipment load: 5.6 kW

6 Roof insulation: the exposed roof of the air-conditioned areas will be insulated

7 Electrical power supply: 415 v/3 ph/50 Hz, AC power supply

8 Humidity control: not considered

9 Glass: SHGC-0.25

10 U-value: 0.57 Btu/hr/sq F

## 3.3 EXPERIMENTAL SETUP

To maintain acceptable indoor air conditions and heat comfort, standard ventilation systems must be designed to amalgamate interior-contaminated air by bringing in clean air from the outdoors. In general, ventilation systems are classified into two types based on their working forces: natural ventilation and forced ventilation. The former is generally used as a backup because of difficulty in creating steady wind pressure. In order to fulfil indoor air quality standards, mechanical ventilation regulation is primarily used to improve the pollutant removal process. Suggested mechanical ventilation solutions may vary from standard air supply techniques, like before-hand amalgamation ventilation, displaced ventilation, subsurface air circulation, and individualised ventilation. The removal effectiveness of internal air contaminants relies on the ventilation or circulation pattern in the context of heat and ventilation.

From the standpoint of ventilation configuration, the core issue in regulating interior biological pollutants via ventilation systems is to utilise a process that completely nullifies and destroys the interior biological pollutant under a specific combined effect of ventilation amount, air flow method, and possibly ultraviolet germicidal irradiation (UVGI). Regulation of ventilation systems in confined and heavily used environments such as medical institutions and public usage zones is desperately needed in light of the current COVID-19 situation. The terms acquired from metrological department of India are as provided in Table 3.1, and further analysis was performed using this data.

The primary requirements of this research comprise three major prediction parameters. Henceforth prime attention here is given to measurement and predication of overall sun irradiation (SR), dry bulb air temperature (DBT) and relative humidity (RH). For this purpose, the compiled data is designed and tested in an intelligent way by incorporating

TABLE 3.1   Details of the Experimental Location

| S.No. | Properties of location | Values |
|---|---|---|
| 1 | Latitude | 28.6139° N |
| 2 | Longitude | 77.2090° E |
| 3 | Elevation | 216 m |
| 4 | Area | 42.7 km² |
| 5 | Weather | 38°C–46°C in summers |
| 6 | Wind | E 11 km/h |
| 7 | Humidity | 40%–50% summers |

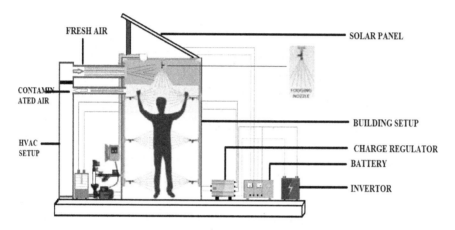

FIGURE 3.1   Experimental setup of the combined HVAC-IoT-UVGI technology

an ANFIS framework, which, besides being energy autonomous, offers a minimum error rate, predicts the three variables mentioned earlier, and provides forecasts of these variables over a user-specified time. SHT2 sensors were used for the measurement of dry bulb air temperature and relative humidity: In both, the two variables are measured by the same device. The experimental setup, which is an integration of HVAC-IoT-UVGI, is shown in Figure 3.1.

## 3.4 RESULTS AND DISCUSSION

Considering the significant cost savings provided by MBPM-UVGI for HVAC management, predictive control necessitates the inclusion of supplementary sensors, presumably for assessing factors linked to room thermal environment, virus identification, and perhaps outside weather elements. All these necessary activities require an Internet of Things network to facilitate the operation. Information regarding the outside weather as well as the internal environment of the rooms under management is needed. As a result, it necessitates the acquisition of the parameters required to define the weather conditions. The rate of infiltration is affected by various elements, including the rigidity of the building's walls, apertures, and doors, as well as the predominant wind speed and pattern. As previously stated, the air change technique or crack approach is used to obtain the infiltration rate.

The infiltration rate using the air change method is given by:

$$V_0 = \frac{(ACH).V}{3600} m^3 / s$$

ACH values typically range from 0.5 ACH for rigid and well-secured structures to around 2.0 ACH for weak and inadequately sealed buildings. The ACH value for contemporary structures might be as low as 0.2 ACH. Thus, an acceptable ACH value must be determined based on the history and state of the structure, from which the infiltrating rate may be computed. The infiltration rate using the crack method is given by:

$$V_0 = A.C.\Delta P^n m^3 / s$$

$A$ determines the impactful leakage area of the crevasses; $C$ corresponds to the flow coefficient which varies depending on the form of crevasse and the pattern of the discharge in the crack; $P$ is the variation between external and internal pressure (Po – Pi); and $n$ is a parameter whose value varies depending on the pattern of the discharge in the crevasse. Table 3.2 is a set of readings used for RSM modelling and finding the relationships between various inputs and outputs.

### (A) Integration of HVAC-UVGI with IoT system

There are several prediction techniques which utilise IoT methods to derive metadata, each of which must be adjusted per the system type. In order to provide feeds for IoT services, an IoT system should keep a list of important devices and critical metadata knowledge about them. It must also be feasible to directly manage these devices, modify operation settings, update software, inquire about their performance, and facilitate monitoring of any fault situations. Plots (.pdf or. png documents with magnification abilities), positions (things that can be described by a GPS unit), entities (collection of sensors and/or actuators), sensors (distinctively labelled equipment that can quantify a particular variable and also has a designated type and characteristics), sensor units and emblems,

TABLE 3.2  Estimated Output Values from Input Variables

| S.No. | DBT (°C) | Solar Radiation (W/m²) | Relative Humidity (%) | Heat Load Efficiency | Ventilation Efficiency |
|---|---|---|---|---|---|
| 1 | 11.98 | 7.38 | 87.47 | 0.66 | 0.84 |
| 2 | 13.19 | 7.02 | 85.44 | 0.68 | 0.81 |
| 3 | 16.52 | 6.66 | 80.43 | 0.69 | 0.77 |
| 4 | 17.31 | 6.21 | 67.20 | 0.72 | 0.77 |
| 5 | 16.41 | 6.75 | 76.27 | 0.69 | 0.77 |
| 6 | 15.99 | 7.47 | 91.73 | 0.65 | 0.78 |
| 7 | 18.42 | 7.29 | 84.69 | 0.65 | 0.74 |
| 8 | 19.69 | 7.38 | 88.53 | 0.66 | 0.71 |
| 9 | 20.85 | 6.93 | 83.73 | 0.68 | 0.69 |
| 10 | 19.74 | 6.84 | 83.09 | 0.68 | 0.71 |
| 11 | 16.52 | 8.82 | 96.85 | 0.61 | 0.77 |
| 12 | 18.63 | 6.66 | 79.47 | 0.69 | 0.73 |
| 13 | 20.27 | 6.39 | 71.79 | 0.71 | 0.71 |
| 14 | 18.89 | 7.47 | 83.63 | 0.65 | 0.73 |
| 15 | 21.96 | 7.38 | 81.49 | 0.66 | 0.67 |
| 16 | 22.85 | 7.29 | 76.16 | 0.66 | 0.65 |
| 17 | 24.28 | 7.02 | 77.65 | 0.68 | 0.62 |
| 18 | 20.74 | 6.84 | 76.69 | 0.68 | 0.69 |
| 19 | 23.96 | 5.85 | 58.99 | 0.73 | 0.63 |
| 20 | 21.96 | 5.67 | 57.71 | 0.74 | 0.67 |
| 21 | 26.71 | 5.31 | 52.80 | 0.75 | 0.59 |
| 22 | 23.38 | 6.12 | 64.85 | 0.72 | 0.62 |
| 23 | 27.23 | 5.04 | 55.89 | 0.75 | 0.59 |
| 24 | 29.56 | 4.68 | 47.89 | 0.77 | 0.56 |
| 25 | 26.81 | 5.31 | 64.00 | 0.75 | 0.60 |
| 26 | 25.81 | 5.58 | 62.51 | 0.73 | 0.63 |
| 27 | 28.18 | 4.77 | 60.27 | 0.77 | 0.58 |
| 28 | 25.65 | 6.66 | 77.23 | 0.69 | 0.63 |
| 29 | 30.08 | 6.03 | 64.75 | 0.72 | 0.55 |
| 30 | 25.76 | 7.38 | 87.25 | 0.66 | 0.63 |
| 31 | 27.76 | 6.12 | 59.63 | 0.72 | 0.59 |
| 32 | 31.93 | 5.58 | 54.29 | 0.73 | 0.53 |

users and organisations, notifications with different kinds and routines, and time modules are all considered by the IoT network. The coordinates of the sensor go into the sensor table, the actuator coordinates into the control values table, and forecast coordinates into the climatic table. The error rates established during RSM separated models are shown in Figure 3.3.

TABLE 3.3    Error Rates Estimated from Separated Models of Heat Load and Ventilation

| Relationship Function | Heat Load Error % | Ventilation Error % |
| --- | --- | --- |
| Triangular | 5.39E-05 | 2.97E-05 |
| Trapezoidal | 7.86E-03 | 1.34E-02 |
| Cubic | 1.65E-04 | 1.57E-04 |
| Gaussian 1 | 1.21E-04 | 2.44E-04 |
| Gaussian 2 | 2.96E-04 | 4.95E-03 |
| Polynomial | 0.00938271 | 0.0196119 |
| Generalised Bell | 2.46E-03 | 6.22E-03 |
| Sigmoidal | 2.46E-03 | 6.22E-03 |

### (B)  Solar energy for UVGI technology

Solar panels were used to power the UVGI lamps in the room setup, thereby saving energy. The total radiation energy was measured by an irradiance meter. The precision and energy usage were also taken into consideration while sorting the variables which impact the whole process. The temperature and relative humidity precision values were found to vary between ±0.2 C and ±2%, respectively. The needed calculations were distributed to the collector or storage nodes in order to save battery energy. The luminous PV panels were then linked to the lithium polymer battery in a later test. An ammeter was used to monitor the current provided to the battery, and a sensor was used to detect the ambient illumination level as solar potential. The PV display, a diode, the ammeter, and the batteries were all linked in a sequence in this manner, making the charging network a complete loop.

$$I_{OUT} = 10\ \mu A \times \log(SR)$$

$I_{OUT}$ is the flow of current recorded at the sensor output, and SR is the associated solar radiance, the formula is the relationship between current and illumination.

### (C)  Impact assessment of solar radiation, humidification, and dry bulb temperature

Also, it has been established viruses spread through a particular individual coughing, sneezing, or talking. These release respiratory droplets rich in water which are breathed out through air passages in the lungs. This is evident, as, with low humidity in the surrounding air, these droplets with a higher humidity percentage often lose water content through mass transfusion, thereby decreasing the overall size of the droplets. Moreover, the water content in the particles plays a vital role in diluting the virus, droplets that are low in water content being highly contagious among humans.

In addition, smaller-sized infectious particles tend to travel further under the circular airflow driven by ventilation within the building. Indoor relative humidity between 50% and 60% is often suggested to reduce any inherent risk for dispersion of airborne-contagious diseases. UV lights deriving energy from solar energy have been employed for sterilisation and disinfection and might be installed in an air-handling unit or directly in the ventilation enclosure with no major impact on air flow pattern circulation. This works by breaking down certain cell structures, thereby disrupting the structure of the RNA of the virus, causing the microorganism to become unstable and eventually die before its multiplication. According to different sizes and shapes of microorganisms (in a particular room setup) that affect their ultraviolet ray absorption rate, the required time to kill each species could vary based on these criteria. Application of proper optimisation and modelling for various performance parameters of UVGI devices in ventilated rooms would aid in establishing a cost-efficient and effective sterilisation process. Conventional HVAC systems are usually operated with extreme ventilation rates and temperature settings, with negligible considerations given to the case of energy efficiency of the entire setup. Thus, to simultaneously maintain a healthy indoor environment and minimise building energy usage, efficiency would be the prime challenge. New and emerging building ventilation technologies may provide active solutions to achieve building energy efficiency and saving comprising advanced ventilation systems, optimised system sensing, monitoring and controlling technologies, and data analytics. Also new designs include computational fluid dynamics (CFD) modelling, which recently has gained wide acceptance for the design and application of advanced ventilation systems. But the CFD concept seems far-fetched due to several implications regarding establishing a practical setup. To realise online control of ventilation, the CFD-based machine learning method employing soft computing technologies could be a potential way, which provides feasible and accurate prediction of air-circulation requirements. Furthermore, inter-disciplinary research on the construction of a viable AI-operated ventilation system should be conducted in the future.

While considering the design of an HVAC system, the prime three factors which impact the entire process of disinfection are solar radiation, humidification, and temperature. Previous studies have highlighted the importance of these parameters, and the ideal conditions established for the spread of pathogens and viruses could be facilitated in cold and dry conditions. This was modelled on RSM, and the equated coefficients are given in Table 3.4:

TABLE 3.4    Regression Model Equations for Heat Load

| Source | DF | Adj SS | Adj MS | F-Value | P-Value |
|---|---|---|---|---|---|
| Model | 9 | 0.048575 | 0.005397 | 207.67 | 0 |
| Linear | 3 | 0.029417 | 0.009806 | 377.28 | 0 |
| DBT | 1 | 0.000029 | 0.000029 | 1.1 | 0.305 |
| SR | 1 | 0.000582 | 0.000582 | 22.39 | 0 |
| RH | 1 | 0.000129 | 0.000129 | 4.97 | 0.036 |
| Square | 3 | 0.000194 | 0.000065 | 2.48 | 0.088 |
| DBT×DBT | 1 | 0.000117 | 0.000117 | 4.5 | 0.045 |
| SR×SR | 1 | 0.000103 | 0.000103 | 3.96 | 0.059 |
| RH×RH | 1 | 0.000051 | 0.000051 | 1.96 | 0.175 |
| 2-Way Interaction | 3 | 0.000241 | 0.00008 | 3.1 | 0.048 |
| DBT×SR | 1 | 0.000002 | 0.000002 | 0.07 | 0.797 |
| DBT×RH | 1 | 0.000016 | 0.000016 | 0.61 | 0.443 |
| SR×RH | 1 | 0.000061 | 0.000061 | 2.35 | 0.14 |
| Error | 22 | 0.000572 | 0.000026 | | |
| Total | 31 | 0.049147 | | | |

The regression equation obtained by modelling is given by:

**Heat Load** = 1.452 – 0.01712 DBT – 0.105 SR – 0.00208 RH + 0.000138 DBT×DBT + 0.01518 SR×SR+ 0.000083 RH×RH + 0.00053 DBT×SR + 0.000094 DBT×RH – 0.00194 SR×RH

The model efficiency obtained was 96%, which is within the acceptable range. Contour plots were plotted for all the variables and are shown in Figures 3.2–3.4.

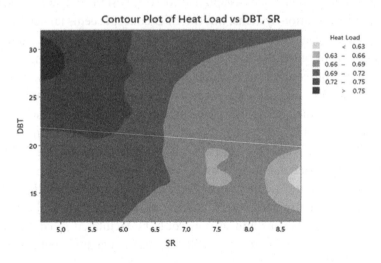

FIGURE 3.2    Contour plot of heat load vs DBT, SR

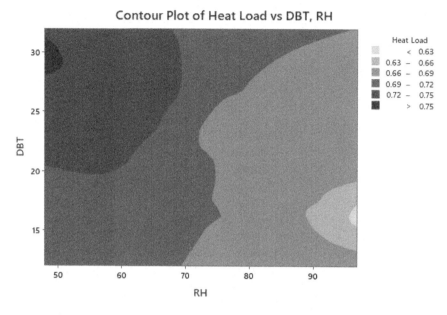

FIGURE 3.3    Contour plot of heat load vs DBT, RH

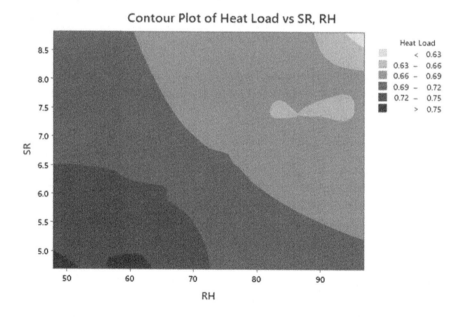

FIGURE 3.4    Contour plot of heat load vs RH, SR

Ventilation efficiency was also evaluated using the RSM method and is presented in Table 3.5.

TABLE 3.5    Regression Model Equations for Ventilation

| Source | DF | Adj SS | Adj MS | F-Value | P-Value |
|---|---|---|---|---|---|
| Model | 9 | 0.205195 | 0.022799 | 182.03 | 0 |
| Linear | 3 | 0.059665 | 0.019888 | 158.79 | 0 |
| DBT | 1 | 0.017665 | 0.017665 | 141.04 | 0 |
| SR | 1 | 0.00003 | 0.00003 | 0.24 | 0.628 |
| RH | 1 | 0.000029 | 0.000029 | 0.23 | 0.638 |
| Square | 3 | 0.000597 | 0.000199 | 1.59 | 0.22 |
| DBT*DBT | 1 | 0.000539 | 0.000539 | 4.3 | 0.05 |
| SR*SR | 1 | 0.000014 | 0.000014 | 0.11 | 0.743 |
| RH*RH | 1 | 0.000001 | 0.000001 | 0.01 | 0.94 |
| 2-Way Interaction | 3 | 0.000295 | 0.000098 | 0.79 | 0.515 |
| DBT*SR | 1 | 0.000017 | 0.000017 | 0.14 | 0.716 |
| DBT*RH | 1 | 0.000085 | 0.000085 | 0.67 | 0.42 |
| SR*RH | 1 | 0.000005 | 0.000005 | 0.04 | 0.846 |
| Error | 22 | 0.002755 | 0.000125 | | |

The regression equation obtained by modelling is given by:

Ventilation = 1.339 – 0.0338 DBT + 0.065 SR – 0.0093 RH + 0.000296 DBT*DBT – 0.0056 SR*SR + 0.000010 RH*RH – 0.00166 DBT*SR + 0.000217 DBT*RH + 0.00055 SR*RH

The model efficiency obtained was 98%, which is again in the acceptable range. Contour plots were plotted for all the variables and are depicted in Figures 3.5–3.7.

## 3.5  CONCLUSION

To improve the quality of air in buildings for a longer span of time, the recent modelled ventilation system can be used because of its technical approach that integrates modern UVGI technology with natural ventilation, and it is also effective in reducing the cost of the overall system. The haphazard design of buildings with impermeable enclosures and inadequate ventilation prevents natural ventilation and necessitates the use of energy-intensive air distribution systems. Enclosed areas are increasingly becoming hotspots for COVID-19 spread since they are enclosed from everywhere preventing air from being transferred from anywhere outside. This analysis furnishes an unorthodox HVAC prototype that enhances the

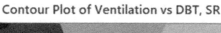

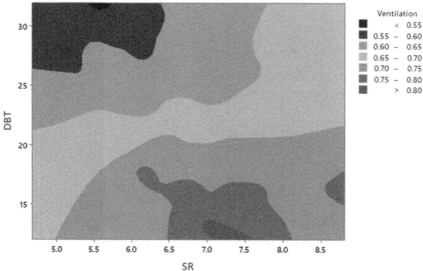

FIGURE 3.5    Contour plot of ventilation vs DBT, SR

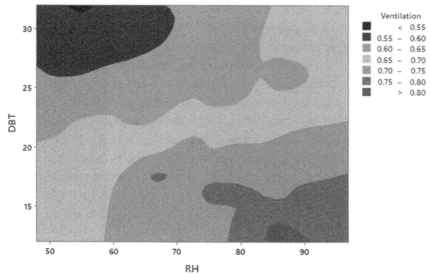

FIGURE 3.6    Contour plot of ventilation vs DBT, RH

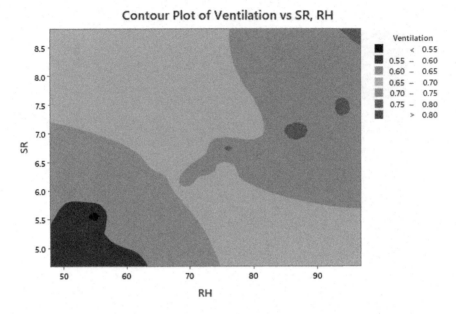

FIGURE 3.7   Contour plot of ventilation vs RH, SR

decoupling of ventilation and thermal control by increasing the outside air supply for typical systems per COVID-19 recommendations. It is quite evident the ventilation and heat load depend on three climatic variables: solar radiation, relative humidity, and DBT. RSM modelling predicted lower error with higher efficiency and can be used to further enhance models for data prediction and elimination of viruses. To reduce the danger of infectious disease transmission within buildings and also to reduce energy usage, this suggested system will work in conjunction with natural air circulation. Our findings demonstrate that boosting outside air in traditional systems may double cooling expenses, whereas enhancing natural ventilation using a UVGI system can bring down the expenses to half in different regions. In 20 major Indian cities, this strategy reduces energy expenditures by 20%–50% while simultaneously improving fresh air intake.

## REFERENCES

[1] R. Pacheco, J. Ordóñez and G. Martínez, "Energy efficient design of building: A review," *Renewable and Sustainable Energy Reviews*, vol. 16, no. 6, pp. 3559–3573, 2012.
[2] S. Zhan and A. Chong, "Building occupancy and energy consumption: Case studies across building types," *Energy and Built Environment*. https://doi.org/10.1016/j.enbenv.2020.08.001.

[3] S. Chen, G. Zhang, X. Xia, S. Setunge and L. Shi, "A review of internal and external influencing factors on energy efficiency design of buildings energy & buildings," *Energy and Buildings*, vol. 216, p. 109944, 2018.

[4] X. Wang, H. Altan and J. Kang, "Parametric study on the performance of green residential buildings in China," *Frontiers of Architectural Research*, vol. 4, pp. 56–67, 2015.

[5] J. Singh, S. S. Mantha and V. M. Phalle, "Analysis of technical and economic electricity saving potential in the urban Indian households," *Sustainable Cities and Society*, vol. 43, pp. 432–442, 2018.

[6] M. A. Cusenza, F. Guarino, S. Longo, M. Mistretta and M. Cellura, "Environmental assessment of 2030 electricity generation scenarios in Sicily: An integrated approach," *Renewable Energy*, vol. 160, pp. 1148–1159, 2020.

[7] Z. Wu, B. Wang and X. Xia, "Large-scale building energy efficiency retrofit: Concept, model and control energy," *Energy*, Elsevier, vol. 109, pp. 456–465, 2016.

[8] S. N. Murray, B. Rocher and D. T. J. Sullivan, "Static simulation: A sufficient modelling technique for retrofit analysis," *Energy and Buildings*, vol. 47, pp. 113–121, 2012.

[9] D. Suzi and G. K. L. Oral, "A study on life cycle assessment of energy retrofit strategies for residential buildings in Turkey," Energy Procedia, vol. 78, pp. 842–847, 2015.

[10] D. Zheng, L. Yu and L. Wang, "Research on large-scale building energy efficiency retrofit based on energy consumption investigation and energy-saving potential analysis," *Journal of Engineering*, vol. 145, no. 6, p. 04019024, 2019.

[11] H. A. Toosi, M. Lavagna, M. F. Leonforte, C. D. Pero and N. Aste, "Life cycle sustainability assessment in building energy retrofitting: A review," *Sustainable Cities and Society*, vol. 60, p. 102248, 2020.

[12] F. Ardente, G. Beccali, M. Cellura and V. Brano, "Life cycle assessment of a solar thermal collector: Sensitivity analysis, energy and environmental balances," *Renewable Energy*, vol. 30, no. 7, pp. 1031–1054, 2005.

[13] B. Griffith, N. Long, P. Torcellini and R. Judkoff, "Assessment of the technical potential for achieving net zero-energy buildings in the commercial sector," *Technical Report NREL/T*, pp. 550–41957, 2007.

[14] J. Kneifel, "Life-cycle carbon and cost analysis of energy efficiency measures in new commercial buildings," *Energy and Buildings*, vol. 42, pp. 333–340, 2010.

[15] Z. Ma, P. Cooper, D. Daly and L. Ledo, "Existing building retrofits: Methodology and state-of-the-art," *Energy Buildings*, vol. 55, pp. 889–902, 2012.

[16] E. Asadi, M. G. D. Silva, C. H. Antunes and L. Dias, "A multi-objective optimization model for building retrofit strategies using TRNSYS simulations," *Gen Opt and MATLAB. Building and Environment*, vol. 56, pp. 370–378, 2012.

[17] T. Zhang, X. Liu, L. Zhang, J. Jiang, M. Zhou and Y. Jiang, "Performance analysis of the air-conditioning system in Xi'an Xianyang International Airport," *Energy Buildings*, vol. 59, pp. 11–20, 2013.

[18] R. Chedwal, J. Mathur, G. D. Agarwal and S. Dhaka, "Energy saving potential through energy conservation building code and advance energy efficiency measures in hotel buildings of Jaipur city, India," *Energy Buildings*, vol. 92, no. 282–295, 2015.

[19] K. Sun and T. Hong, "A simulation approach to estimate energy savings potential of occupant behaviour measures," *Energy and Buildings*, vol. 136, pp. 43–62, 2017.

[20] Y. Liu, T. Liu, S. Ye and Y. Liu, "Cost-benefit analysis for energy efficiency retrofit of existing buildings: A case study in China," *Journal of Cleaner Production*, vol. 177, no. 493–506, 2018.

# CHAPTER 4

# Case Study on Modernization of a Micro-Grid and Its Performance Analysis Employing Solar PV Units

Shiva Pujan Jaiswal, Vikas Singh Bhadoria,
Ranjeeta Singh, Vivek Shrivastava, A. Ambikapathy

## CONTENTS

## 4.1 INTRODUCTION

A micro-grid is a collection of interconnected loads and distributed energy resources (distributed generation and energy storage) that operate as a single controlled entity within a well-defined electrical area. It is a small-scale power grid that may function independently or in synchronization with the local electrical grid. A small-scale localized station with its own power resources, loads, and specified limits qualifies as a micro-grid [1]. It can be used as a backup power source or to supplement the main power grid during peak demand periods. Micro-grids frequently use several energy sources to incorporate renewable energy [2]. Other goals include cost reduction and increased dependability. Micro-grids use a modular architecture that may make the main grid less vulnerable to localized faults. Micro-grids can also be utilized piecemeal to gradually upgrade the grid schematic due to its modularity. Distributed, scattered, decentralized, district, and embedded energy generation are terms used to describe the use of micro-grids [3]. The grid is a system that connects houses, businesses, and other structures to centralized power sources, allowing us to utilize appliances, heating systems, and devices. However, because the grid is so interconnected, when a section of it has to be fixed, every section is affected. This is where a micro-grid might assist [4]. A micro-grid can run while linked to the grid, but it can also disconnect and operate independently, utilizing local energy generation in the event of a disaster, such as a storm or power outage, or for other reasons.

One of the major components of a modern micro-grid is incorporating automation at every stage of the power system, namely generation, transmission, and distribution. Distribution automation refers to automation of distribution sub-stations, feeders, and major loads. It includes monitoring, control, and protection of the distribution sub-station and remote control of feeders. The modern technique for such distribution control is supervisory control and data acquisition (SCADA) [5]. A SCADA system consists of a number of remote terminal units (RTUs) and a master terminal unit (MTU). An RTU provides an interface to different field devices through its digital inputs (DIs) and digital outputs (DOs).

A SCADA system for a micro-grid is designed and implemented in [6]. Front-end processing modules (RTUs), supervisory control modules (MTUs), and communication gateway modules are all part of the architecture. The front-end processing modules and field devices connect via a controller area network (CAN) bus, while the front-end processing modules and supervisory module communicate via Ethernet. The design of a laboratory-based SCADA system is presented in [7–8]. [9] presents a concept for developing a system for operating a substation's primary circuit breaker and collecting, displaying, and storing electrical characteristics for future reference. The receiver receives the status of the substation via a virtual proxy network (VPN).

In addition to the fundamental measurement of energy, a variety of characteristics linked to power quality are monitored in current electric substations of power distribution networks and recent micro-grids. Smart multifunction metres (SMFMs) are used instead of ordinary energy metres to achieve this. SMFMs have the capacity to communicate as well as to measure up to 100 different electrical parameters of the microgrid as discussed in reference [10]. It presents smart metres as a critical component of smart grids and emphasises data collection, processing, and interpretation. [11] explains why smart multifunction metres are preferred over traditional digital metres. The methods and approaches utilized in the construction of smart metres are detailed in [12]. Data from smart energy metres must be collected remotely and automatically at a central location, which is usually the utility's headquarters or control room (distribution company). Data collection from smart metres may be done in a variety of ways, including employing wireless or wired connection technologies. According to [13], smart energy metres employ two forms of wireless communication, one for short distances and the other for long distances. The energy consumption is sent to a central node across short distances using an RF transmitter included in the energy metre. The measurements from the central node are sent to the utility office through a global system for mobile (GSM) module over great distances. When data is collected from metres set in consumers' homes with the primary goal of creating power bills for payment by consumers, wireless communication technologies can perform well. However, due to the strong electro-magnetic interference (EMI) created by high voltages (11 or 33 kV) and high currents (typically several hundred amperes) present in the distribution substation, wireless communication technologies will not work satisfactorily when data is to be acquired from SMFMs installed in the substation. Because it is intended for the functioning of the distribution system and future planning, error-free data exchange is both necessary and critical. The usage of wired communication technology is the answer.

In this chapter, the automation of a practical micro-grid is carried out. A system is designed to collect energy generation data from PV units installed on the roofs of ten different buildings of an institute. The data kWh of generation is collected at RTUs, then this data is sent to the MTU. A local area network (LAN) is used for MTU-RTU communication. Each RTU uses an Allen Bradley Micro Logix 1400 programmable logic controller (PLC). This chapter presents a single-line diagram of the micro-grid, hardware design of the RTU located in one building, and ladder programs for the PLC of this RTU. The design also includes the RS-485 network used for communication between the RTU and SMFMs. The program for the PLC was developed using the ladder programming language on an RS Logix 500 Emulator. After collecting the monthly data of kWh generation by the different units of the PV plant, a comparison of the power generation is carried out to analyze their performance in the same environment. The financial loss is also calculated due to inferior performance of a few units with respect to the best performer. The chapter is divided into four sections; the first section presents the previous work of other researchers and provides an introduction to the chapter. The second section presents a description of the micro-grid and its automation. The third section presents the analysis of monthly energy generation by the ten solar PV units. The fourth section presents the conclusion.

## 4.2 MODERNIZATION OF A MICRO-GRID

Modernization of micro-grids is essential to monitor and control the operation of the system. Automation of a micro-grid provides numerous advantages such as online monitoring, data collection, opening and reclosing of circuit breakers, and so on. A SCADA-based PLC system is developed for automation of the micro-grid.

### 4.2.1 Details of Micro-Grid

In this chapter, a Sharda University practical micro-grid is studied. It consists of two 33/11-kV transformers with an aggregate capacity of 8 MVA, six 11/0.433-kV transformers with an aggregate capacity of 10 MVA, one 33-kV feeder, three 11-kV feeders, 177 415-V main feeders with an aggregate capacity of 10 MW, and captive diesel generators. Noida Power Corporation Ltd. (NPCL) is the primary source of supplies. The 33-kV power supply was provided by NPCL and was then stepped down to 11 kV and distributed to three substations: ESS1, ESS2, and ESS3. All three substations are also linked. The voltage is stepped down to 415 V at the substation and distributed to the loads of various buildings. Initially, it was merely a distribution system, but today the ten buildings have their own rooftop solar PV power plants. It produces around 1 million kWh per year. This generated energy is sent into the building's main distribution panel, which is also where the rooftop solar power plant is located. The buildings' energy is supplied by two different sources: NPCL and rooftop PV units. Because the demand for electricity on campus is typically more than 1 MW, batteries are not required to store power because it may be consumed directly. Net metering is also possible, allowing the institution to provide back to the public grid. Figure 4.1 depicts the system as a single-line diagram.

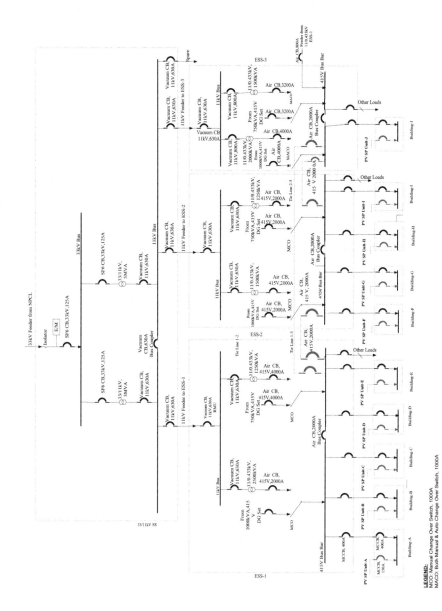

FIGURE 4.1    Single-line diagram of micro-grid

To make it smart, this micro-grid is being modernized and enhanced. A SCADA system will be used to control this micro-grid from a central location. The design and programming of an RTU for supervisory control of one of the institution's four electric substations and buildings is presented in this chapter.

### 4.2.2 Proposed Supervisory Control System

A remote terminal unit, a master terminal unit, field devices (FDs), an MTU-RTU communication system, and an RTU-FD communication system make up the supervisory management of an electric substation. The next subsections go through each of these components in detail. Figure 4.2 depicts the suggested scheme for installing 18 RTUs in various locations. The institute has its own campus-wide LAN, which has been used to connect MTUs and RTUs.

*RTU-Field Device Communication System:* Individual copper wires connect the RTU to non-smart field equipment like circuit breakers and tap changers. The smart multi-function metres in the substation, on the

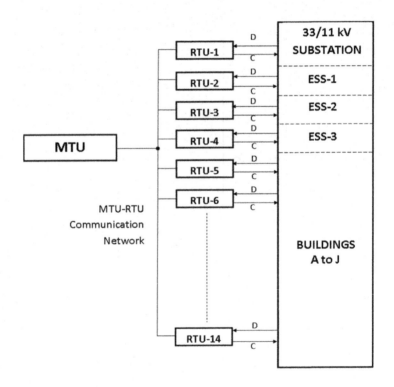

FIGURE 4.2   Proposed scheme of RTU-MTU communication network

other hand, are connected through a personal area network (PAN) that terminates at the RTU's PLC. Through the PAN, the RTU connects with the smart metres. The smart metres and PLC in the PAN were connected via a shielded twisted pair (STP) connection. The specifics of RTU-2 for ESS-1 are presented in this chapter. The MTU is positioned in the control room and has two primary functions: (a) to continuously monitor the whole power distribution system/process and (b) to provide control directives as needed. The MTU sends these directives to the RTU. These control messages are converted into control signals by the RTU and sent to the appropriate field devices, such as circuit breakers and tap changers [4].

*Single Line Diagram of ESS-1:* The single line schematic of the electric substation, ESS-1, is shown in Figure 4.3. It's a distribution substation that transforms 11 kV into 415 V. It can power up to 26 load points. The Schneider EM 6400 and EM 6436 SMFMs are employed in this substation. Its functioning is controlled by eight segment LED displays, an analogue load bar, various LED indicators, and five smart keys. The metres are able to communicate and can measure over 100 different values. The EM 6400 and EM 6436 power terminals are situated on the back panel. There are 14 terminals total, 7 on each side, including 6 terminals for current, 4 for voltage, 2 for auxiliary power supply, and 2 for the RS 485 communication port [14].

*Hardware Design of Remote Terminal Unit:* The RTU's primary component is a PLC. Figure 4.4 depicts the PLC's digital inputs and their connections. One of the RTU's most significant duties is to use the status information it collects to identify the status of two state objects. The auxiliary coil in moulded case circuit breakers (MCCBs) and air circuit breakers (ACBs) provides this information [15]. Digital outputs are utilized to deliver essential control signals to MCCBs and ACBs in the same way. There are 12 digital outputs on the PLC utilized in the substation. These DOs are connected to the tip coils and motor-based closing mechanisms of the MCCBs and ACBs through relays, as illustrated in Figure 4.8, to open and shut them, respectively.

Since the PLC's DIs and DOs are insufficient to link all of the field devices, such as MCCBs and ACBs, two digital input expansion units and four digital output expansion units were installed.

*Ladder Programming of RTU:* On the Allen Bradley RS Logix 500 Emulator, a program for the RTU's PLC was developed using the ladder programming language. Inputting status information, outputting control signals, and reading SMFMs are the three main functions of the software.

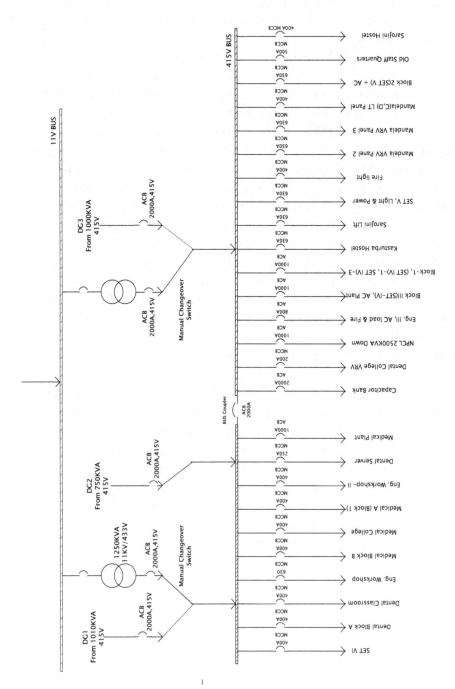

FIGURE 4.3    Single-line diagram of ESS-1

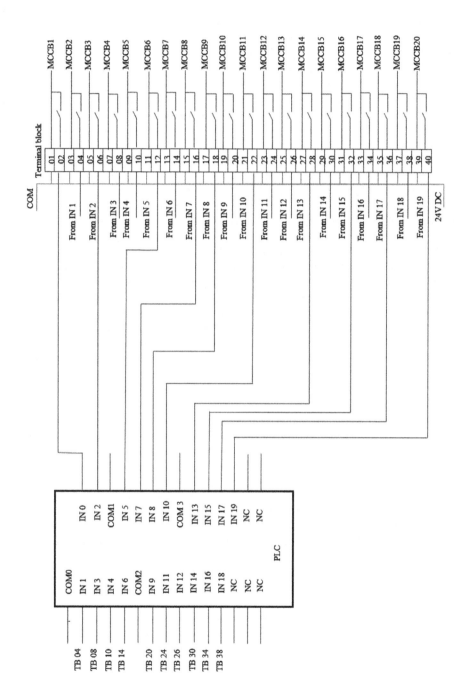

FIGURE 4.4  Connection diagram of digital input terminals of PLC

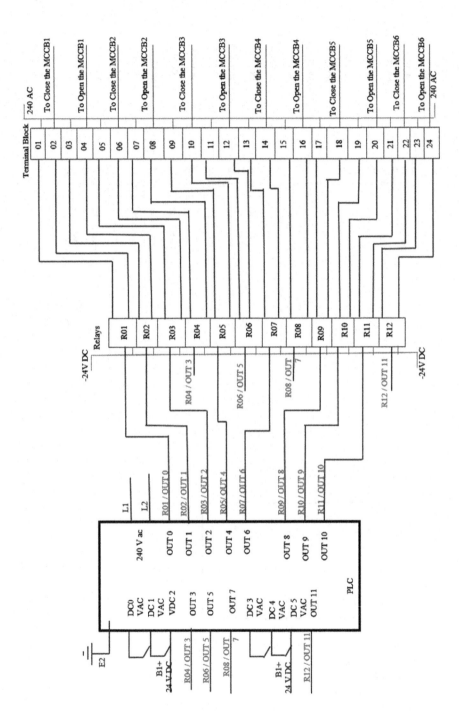

FIGURE 4.5  Connection diagram of digital output terminals of PLC

A ladder program was also created for MCCB status monitoring and control, plus ladder software for obtaining data from smart metres. Every second, the PLC sends a message to each metre, instructing it to submit data. The program in Rung 0 is in charge of this. The value of the frequency supplied by the metres is obtained by Rung 1 and placed in N7:15, which is a register designed to hold integer values, but the value is in decimals. As a result, CPW (copy word) is utilized, and the value is placed in F8:0, the register for floating-point numbers.

*Message Frame Format for Metres:* Figure 4.6 depicts the message format. The PLC sends this message to the smart metres in order to obtain the values of all electrical parameters from the latter. The PLC has three channels, the first of which is the setup for Modbus communication. The message indicates that it will access 100 bytes of data from Modbus address 43901 and save it in the PLC's N7:1 register.

The values of various parameters as obtained from one SMFM are shown in Figure 4.6. The data from the metre is saved in the F8 register, which is designed to hold floating-point numbers. F8:0 stores the frequency value, F8:1, F8:2, and F8:3 store the values of phase voltage, F8:4, F8:5, and F8:6 are used to store the values of line voltage, and F8:11 stores power factor.

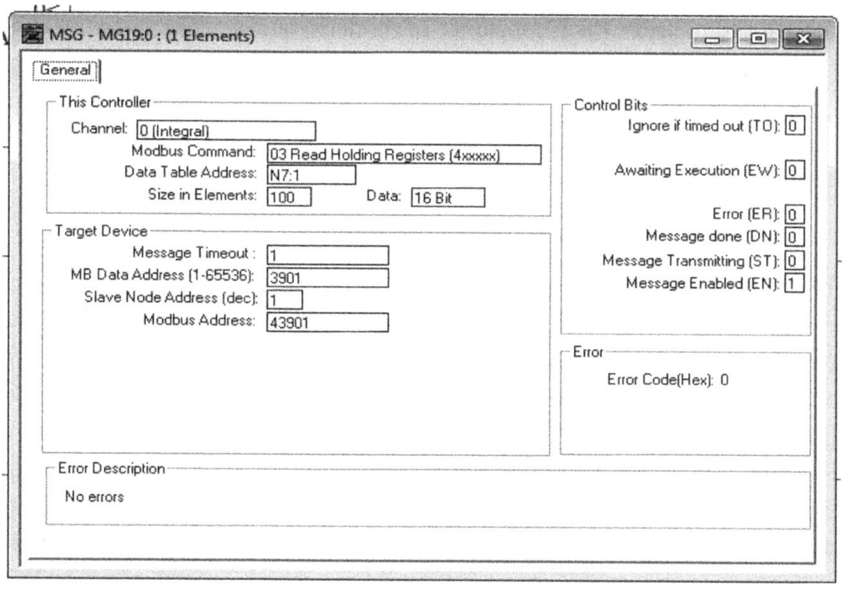

FIGURE 4.6   Message frame format

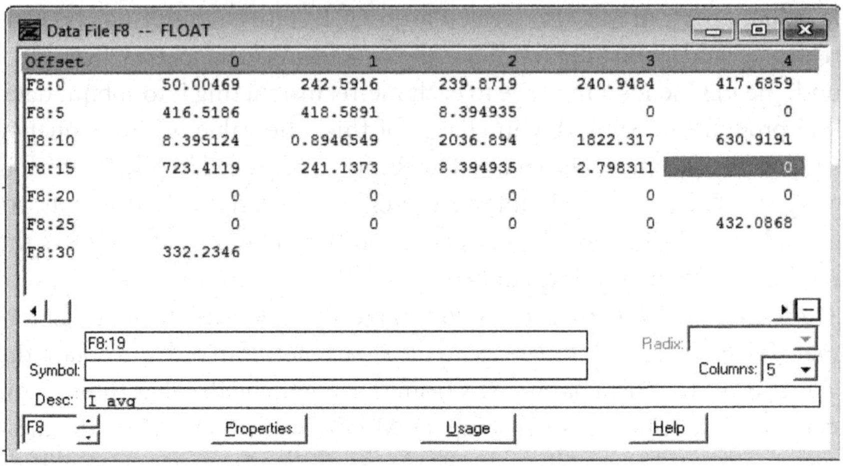

FIGURE 4.7   Floating register of PLC

## 4.3  ANALYSIS OF SOLAR ENERGY GENERATION DATA

In the previous section, a design of data collection using a SCADA system is developed to monitor and control. RTU and MTU communication is also developed to observe the power generation by the ten units located on the rooftops of buildings on the same campus. The monthly energy generated by the units is given in Table 4.1. The kWh energy generated in the year 2019 is analyzed to prove the importance of modernization of the micro-grid.

The monthly kWh generated is given in Table 4.1; the comparison of performance in this form of data is meaningless because the capacity of each unit is different [16]. To normalize the data, the monthly energy generation is divided by the peak plant capacity.

$$kWh/kWp = kWh \text{ generated in month/peak plant capacity} \quad (1)$$

The kWh per kWp is given in Table 4.2. This table shows the cooperative energy values generated per kilowatt installation by each unit. The comparison of kWh generated per kWp installed capacity in each month of 2019 is shown in Figure 4.8(a) for January. The best performer is Building C, and the worst performer is Building F in February, shown in Figure 4.8(b). The performance of each building in March is presented in Figure 4.8(c). In April, Buildings F and J showed poor performance, as shown in Figure 4.8(d), whereas in May, Building J showed superior performance, as shown in Figure 4.8(e). Buildings E and F had the lowest kWh per kWp in June, as shown in Figure 4.8(f). The kWh per kWp for July is

TABLE 4.1  Energy Generated in Each Month of 2019

| | kWp | Jan | Feb. | Mar | Apr | May | Jun | Jul | Aug | Sept | Oct | Nov | Dec | Total |
|---|---|---|---|---|---|---|---|---|---|---|---|---|---|---|
| A | 100.4 | 8855 | 8810 | 13369 | 14020 | 14706 | 13636 | 10654 | 10694 | 10820 | 10743 | 7795 | 6677 | 130779 |
| B | 117.8 | 9961 | 9486 | 15600 | 14854 | 15909 | 14204 | 10669 | 11667 | 10803 | 11739 | 9199 | 6988 | 141079 |
| C | 22.32 | 2067 | 1950 | 2932 | 2907 | 2968 | 2633 | 2038 | 2130 | 2129 | 2421 | 1751 | 1491 | 27417 |
| D | 102.9 | 8722 | 8750 | 12794 | 13225 | 14249 | 12877 | 9976 | 10052 | 10102 | 10052 | 7303 | 6491 | 124593 |
| E | 83.7 | 5551 | 5843 | 7693 | 9739 | 10655 | 9619 | 7291 | 7622 | 6851 | 6958 | 4849 | 4109 | 86780 |
| F | 116.6 | 9007 | 8730 | 13046 | 13149 | 14006 | 12671 | 9842 | 10390 | 10172 | 10954 | 7960 | 6719 | 126646 |
| G | 100.4 | 9150 | 9005 | 13242 | 13507 | 14854 | 13360 | 10106 | 10708 | 9863 | 10020 | 7966 | 6729 | 128510 |
| H | 23.56 | 1947 | 1907 | 2853 | 2966 | 3211 | 2915 | 2290 | 2358 | 2330 | 2359 | 1699 | 1437 | 28272 |
| I | 22.32 | 1848 | 1796 | 2638 | 2803 | 2955 | 2683 | 2056 | 2150 | 2114 | 2189 | 1605 | 1363 | 26200 |
| J | 111.6 | 8869 | 9365 | 14619 | 11352 | 18104 | 16535 | 12919 | 13149 | 12112 | 11662 | 8305 | 7230 | 144221 |
| Total | 801.7 | 65977 | 65642 | 98786 | 98522 | 111617 | 101133 | 77841 | 80920 | 77296 | 79097 | 58432 | 49234 | 964497 |

shown in Figure 4.8(g). The performance of each building in terms of kWh per kWp for August is shown in Figure 4.8(h). The ratio of kWh to kWp for September is shown in Figure 8(i). The kWh per kWp for each building for October is shown in Figure 4.8(j). In November and December, Building G had the highest kWh per kWp, as shown in Figures 4.8(k) and 4.8(l), respectively.

The monthly loss of energy generated or annual monetary loss is estimated by comparing the average energy generated to two units of the worst performer in each month.

$$LGWPU_1 = (kWh/kWp \ of \ BPU - kWh/kWp \ of \ WUP_1) \times kWp \ of \ WUP_1 \quad (2)$$
$$LGWPU_2 = (kWh/kWp \ of \ BPU - kWh/kWp \ of \ WUP_2) \times kWp \ of \ WUP_2 \quad (3)$$
$$TLG_M = LGWPU_1 + LGWPU_2 \quad (4)$$

$BPU$ = best-performing unit
$WPU_1$ = first worst-performing unit
$LGWPU_1$ = loss of energy generation because of first worst-performing unit
$WPU_2$ = second worst-performing unit
$LGWPU_2$ = loss of energy generation because of second worst-performing unit
$TLG_M$ = total loss of kWh generation in each month

The total loss of kWh generation is the sum of the loss of kWh generated in each month because of two worst performers in the system [17]. The total monetary loss is equal to the cost of energy per unit multiplied by the total loss of energy generated in year 2019. The monthly ratios of kWh to kWp of each unit of PV plant are given in Table 4.2.

An estimation of loss of energy generation due to the inferior performance of two units is given in Table 4.3. The total annual loss of energy generation due to the poor performance of two units is equal to 61,749.32 kWh. The actual units generated in 2019 are 964,497 kWh, whereas the possible generation in 2019 is 1,026,246.32 kWh. The monthly gap between the actual and possible generation is shown in Figure 4.9. The total loss of generation in 2019 is demonstrated in Figure 4.10.

In the previous section, an analysis was carried out for the unit of energy generated in each month of 2019. In this part, the annual analysis is carried out. The annual unit/kWp generated by each building is given in

TABLE 4.2  Monthly kWh/kWp of PV Units

|   | kWp | January | February | March | April | May | June | July | August | September | October | November | December | Year |
|---|---|---|---|---|---|---|---|---|---|---|---|---|---|---|
| A | 100.40 | 88.20 | 87.75 | 133.16 | 139.64 | 146.47 | 135.82 | 106.12 | 106.51 | 107.77 | 107.00 | 77.64 | 66.50 | 1302.58 |
| B | 117.80 | 84.56 | 80.53 | 132.43 | 126.10 | 135.05 | 120.58 | 90.57 | 99.04 | 91.71 | 99.65 | 78.09 | 59.32 | 1197.61 |
| C | 22.32 | 92.61 | 87.37 | 131.36 | 130.24 | 132.97 | 117.97 | 91.31 | 95.43 | 95.39 | 108.47 | 78.45 | 66.80 | 1228.36 |
| D | 102.90 | 84.76 | 85.03 | 124.33 | 128.52 | 138.47 | 125.14 | 96.95 | 97.69 | 98.17 | 97.69 | 70.97 | 63.08 | 1210.82 |
| E | 83.70 | 66.32 | 69.81 | 91.91 | 116.36 | 127.30 | 114.92 | 87.11 | 91.06 | 81.85 | 83.13 | 57.93 | 49.09 | 1036.80 |
| F | 116.60 | 77.25 | 74.87 | 111.89 | 112.77 | 120.12 | 108.67 | 84.41 | 89.11 | 87.24 | 93.95 | 68.27 | 57.62 | 1086.16 |
| G | 100.40 | 91.14 | 89.69 | 131.89 | 134.53 | 147.95 | 133.07 | 100.66 | 106.65 | 98.24 | 99.80 | 79.34 | 67.02 | 1279.98 |
| H | 23.56 | 82.64 | 80.94 | 121.10 | 125.89 | 136.29 | 123.73 | 97.20 | 100.08 | 98.90 | 100.13 | 72.11 | 60.99 | 1200.00 |
| I | 22.32 | 82.80 | 80.47 | 118.19 | 125.58 | 132.39 | 120.21 | 92.11 | 96.33 | 94.71 | 98.07 | 71.91 | 61.07 | 1173.84 |
| J | 111.60 | 79.47 | 83.92 | 130.99 | 101.72 | 162.22 | 148.16 | 115.76 | 117.82 | 108.53 | 104.50 | 74.42 | 64.78 | 1292.30 |
| Total | 801.70 | 82.30 | 81.88 | 123.22 | 122.89 | 139.23 | 126.15 | 97.09 | 100.94 | 96.42 | 98.66 | 72.89 | 61.41 | 1203.06 |

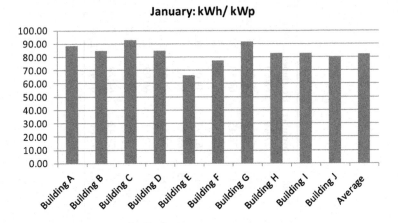

FIGURE 4.8(a)   Performance for January

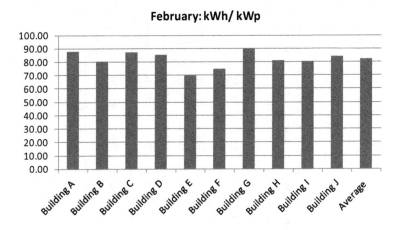

FIGURE 4.8(b)   Performance for February

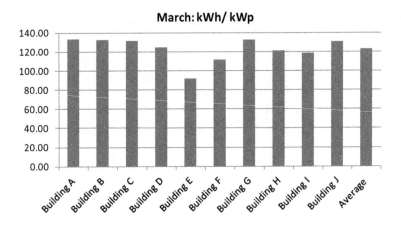

FIGURE 4.8(c)   Performance for March

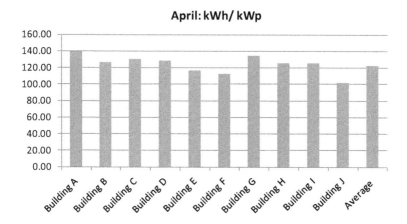

FIGURE 4.8(d)   Performance for April

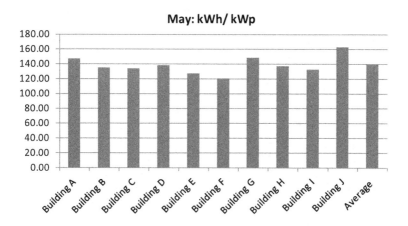

FIGURE 4.8(e)   Performance for May

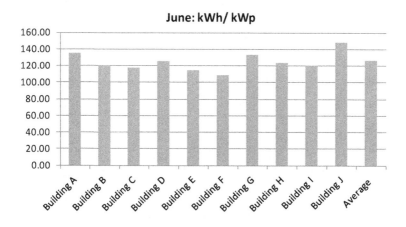

FIGURE 4.8(f)   Performance for June

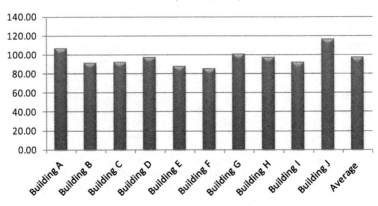

FIGURE 4.8(g)   Performance for July

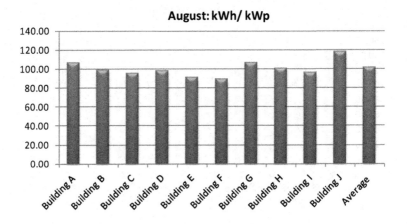

FIGURE 4.8(h)   Performance for August

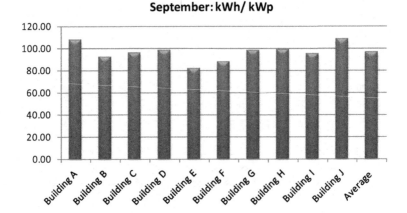

FIGURE 4.8(i)   Performance for September

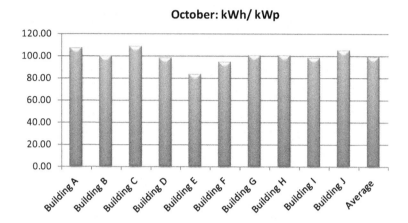

FIGURE 4.8(j)  Performance for October

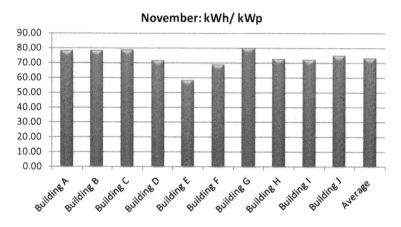

FIGURE 4.8(k)  Performance for November

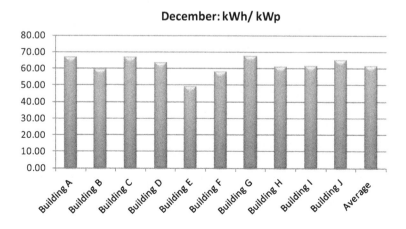

FIGURE 4.8(l)  Performance for December

TABLE 4.3    Monthly Loss of kWh Due to Inferior Performance of Two Units

| Month | BPU | $WPU_1$ | $LGWPU_1$ | $WPU_2$ | $LGWPU_2$ | $TLG_M$ | Actual kWh Generation | Possible Generation |
|---|---|---|---|---|---|---|---|---|
| Jan | C | F | 1790.98 | E | 2200.47 | 3991.45 | 65977 | 69968.45 |
| Feb | G | F | 1728.01 | E | 1663.96 | 3391.97 | 65642 | 69033.97 |
| Mar | A | F | 2480.08 | E | 3452.63 | 5932.71 | 98786 | 104718.71 |
| Apr | A | F | 3133.04 | J | 4231.87 | 7364.91 | 98522 | 105886.91 |
| May | J | E | 2922.80 | F | 4908.86 | 7831.66 | 111617 | 119448.66 |
| Jun | J | E | 2782.19 | F | 4604.53 | 7386.72 | 101133 | 108519.72 |
| Jul | J | E | 2398.01 | F | 3655.41 | 6053.42 | 77841 | 83894.42 |
| Aug | J | E | 2239.81 | F | 3347.59 | 5587.40 | 80920 | 86507.40 |
| Sept | J | F | 2482.41 | E | 2233.12 | 4715.53 | 77296 | 82011.53 |
| Oct | C | F | 1693.03 | E | 2120.96 | 3813.99 | 79097 | 82910.99 |
| Nov | G | F | 1290.76 | E | 1792.02 | 3082.78 | 58432 | 61514.78 |
| Dec | G | F | 1096.04 | E | 1500.74 | 2596.78 | 49234 | 51830.78 |
| Total | | | 26037.17 | | 35712.15 | 61749.32 | 964497 | 1026246.32 |

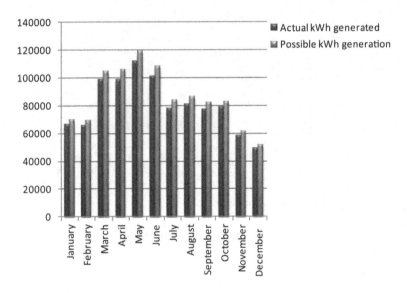

FIGURE 4.9    Gap between the actual generation and possible generation (monthly)

Table 4.4, and the same is shown in Figure 4.11. A comparison of average and expected energy generation is shown in Figure 4.12.

## Possible reasons for low performance of a few PV units under the same conditions [18–19]

1  Improper maintenance, cleaning, or cooling of solar panels

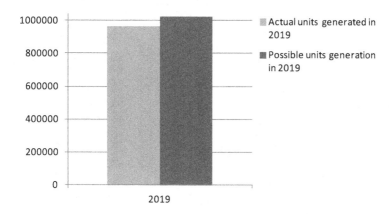

FIGURE 4.10   Total loss of generation in 2019

TABLE 4.4   Annual kWh/kWp of PV Units

| Building | Annual kWh/kWp |
|---|---|
| Building A | 1302.58 |
| Building B | 1197.61 |
| Building C | 1228.36 |
| Building D | 1210.82 |
| Building E | 1036.80 |
| Building F | 1086.16 |
| Building G | 1279.98 |
| Building H | 1200.00 |
| Building I | 1173.84 |
| Building J | 1292.30 |
| Total | 1203.06 |

TABLE 4.5   Annual Loss of kWh Due to Inferior Performance of Two Units

| Year | BPU | $WPU_1$ | $LGWPU_1$ | $WPU_2$ | $LGWPU_2$ | $TLG_A$ | Actual kWh generated | Expected kWh generation |
|---|---|---|---|---|---|---|---|---|
| 2019 | A | F | 25234.57 | E | 22245.79 | 47480.36 | 964497 | 1011977.36 |

2  Partial shadow of tree or other structure

3  Inappropriate MPPT settings

4  Mismatch between generation and demand

5  Low gap between roof and solar panel due to which convective air may not flow to cool them

6  Supporting structure may be dark coloured so that heat absorption will be high

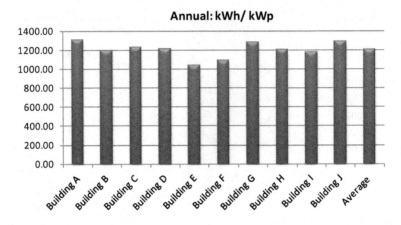

FIGURE 4.11   Performance in 2019

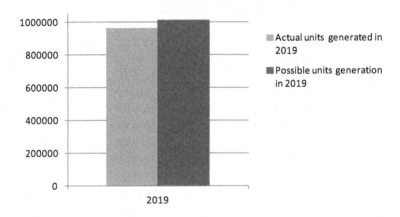

FIGURE 4.12   Loss of generation

## 4.4 CONCLUSION

In order for a plant or process to operate reliably and effectively, supervision and control are critical. These allow the process operator in the control room to view what is going on in the process. The supervisory control system's master unit continually monitors the plant through RTUs, creates alerts in the event of any abnormal situations, and sends required control commands to the plant via RTUs. This chapter presents the general design features of the supervisory control and data collecting system. The hardware modules of DIs and DOs of the PLC of RTU of ESS-1, as well as their extension units, have been described in depth. The ladder programming language was used to create and test the PLC program on the Allen Bradley Micro Logix 1400.

An analysis of the energy generation by ten PV units at an institute has been carried out. The data and results show that the different units did not produce same amount of energy, even when they were installed in the same location. The reason for this dissimilar performance of units is yet to be examined. In this chapter, the analysis was carried out using two approaches: (a) on a monthly basis and (b) annually. The calculated loss of energy generation is more for the monthly approach than the annual approach because the time period is shorter.

Finally, automation of the micro-grid is essential to discover the faults, malfunctioning, and slip in performance of the equipment and to implement control actions to improve the performance of the micro-grid.

## REFERENCES

[1] O. Usta, N. Mahdavi Tabatabaei, E. Kabalci and N. Bizon, "Microgrid protection and automations and architectures, control and protection methods," in *Power Systems*, Cham, 2020. https://doi.org/10.1007/978-3-030-23723-3_26.

[2] R. Srivastava, A. N. Tiwari and V. K. Giri, "An overview on performance of PV plants commissioned at different places in the world," *Energy for Sustainable Development*, vol. 54, pp. 51–59, February 2020. https://doi.org/10.1016/j.esd.2019.10.004.

[3] S. P. Jaiswal, V. S. Bhadoria, V. Shrivastava and N. S. Pal, "Reliability improvement of distribution system by optimal sitting and sizing of disperse generation," *International Journal of Reliability, Quality and Safety Engineering*, 2017. https://doi.org/10.1142/S021853931740006X.

[4] S. P. Jaiswal, V. Srivastava and D. K. Palwalia, "Opportunities and challenges of PV technology in power system," *Materials Today: Proceedings*, 2020. doi:10.1016/j.matpr.2020.01.269.

[5] S. Tan, D. De, W. Song, J. Yang and S. K. Das, "Survey of security advances in smart grid: A data driven approach," *IEEE Commun Surv Tutorials*, vol. 19, no. 1, pp. 397–422, 2017. https://doi.org/10.1109/COMST.2016.2616442

[6] Y. Chen and W. Pie, "Design and implementation of SCADA system for micro-grid," *Information Technology Journal*, pp. 8049–8057, 2013.

[7] H. K. Verma, "Chapter-2 hardware of supervisory control & data acquisition system," *Smart Grid e-monogram at www.profhkverma.info*, vol. 1, pp. 01–19, 2014.

[8] Y. Rangelov, A. Avramov and N. Nikolaev, "Design and construction of laboratory SCADA system," in *International School Conference on Information Communication and Energy System and Technology*, pp. 300–303, Sofia, Bulgaria, 2015.

[9] J. Jose, C. Varghese, A. Abraham, J. Joy and A. Koilraj, "Substation automation system for energy monitoring and control using SCADA," *International Journal of Recent Trends in Engineering & Research*, pp. 32–38, 2017.

[10] D. Alahakoon and X. Yu, "Smart electricity meter data intelligence for future energy systems: A survey," *IEEE Transactions on Industrial Informatics*, pp. 1–12, 2015.

[11] H. R. Zala and V. C. Pandya, "Energy meter data acquisition system with wireless communication for smart metering application," *International Journal of Engineering Research and Technology*, pp. 1590–1595, 2014.

[12] A. Khadar, J. A. Khan and M. S. Nagaraj, "Research advancements towards in existing smart metering over smart grid," *International Journal of Advanced Computer Science and Applications*, pp. 84–92, 2017.

[13] Procom: ACE Series Multifunction Panel Meter User Manual, 2014.

[14] Schneider Electric: Conzerv EM6400 Series Power Meters User Manual, 2015.

[15] A. R. Adly, Z. M. Ali, A. M. Abdel-hamed et al., "Enhancing the performance of directional relay using a positive-sequence superimposed component," *Electrical Engineering*, vol. 102, pp. 591–609, 2020. https://doi.org/10.1007/s00202-019-00896-5

[16] S. P. Jaiswal and S. Singh, "Economic viability solar PV power plant in distribution system," *IOP Conference Series: Materials Science and Engineering*, vol. 594, p. 12010, 2019. doi:10.1088/1757-899X/594/1/012010.

[17] S. P. Jaiswal, S. R. Ramavt, N. Goel and V. Shrivastava, "Optimal location and sizing of DG in distribution system and its cost-benefit analysis," *Advances in Intelligent Systems and Computing*. doi:10.1007/978-981-13-1819-1_11.

[18] H. I. Daga and M. S. Bukerb, "Performance evaluation and degradation assessment of crystalline silicon based photovoltaic rooftop technologies under outdoor conditions," *Renewable Energy*, vol. 156, pp. 1292–1300, August 2020. https://doi.org/10.1016/j.renene.2019.11.141.

[19] D. Yadav, N. Singh and V. S. Bhadoria, "Comparison of MPPT algorithms in stand-alone photovoltaic (PV) system on resistive load," in *Machine Intelligence and Smart Systems. Algorithms for Intelligent Systems*. Singapore, 2021. https://doi.org/10.1007/978-981-33-4893-6_34

CHAPTER **5**

# Harnessing Energy for Implantable Biomedical Instruments with IoT Networks

*Implementation and Challenges*

Mahak Narang, Ankit Gambhir, Dr. Mandeep Singh

## CONTENTS

## 5.1 INTRODUCTION

There are various terms mentioned in the literature for energy harvesting, like power scavenging, energy scavenging or ambient power. It is a method for scavenging electrical energy from environmental sources like thermal energy, solar power, wind energy, kinetic energy and many more [1]. The popularity of energy harvesting is growing at a high rate. It is being used as a replacement source for batteries in wireless autonomous devices. The main factor for this replacement is that batteries contain a finite amount of

DOI: 10.1201/9781003218760-5

energy and require periodic replacement or recharge. In addition to this, the disposal of batteries is not environmentally friendly [2].

In biomedical applications, most implantable devices are powered using wires or batteries. These energy sources for biomedical devices have certain hazards to patients like skin infections, chemical side effects due to batteries and many more. In addition to these implantable biomedical devices, using batteries comes with several other problems like a limited lifetime and large size. To overcome all these difficulties, researchers have established various energy-harvesting methods to power implantable biomedical devices [3].

This chapter is a review of all the different energy-harvesting methods used for powering implantable biomedical devices. The environment and humans themselves are the two main sources of harvesting energy for implantable biomedical devices. The purpose of this review is to cover all the methods related to human sources used to date for energy harvesting in biomedical implantable devices. This will provide readers a thorough background to deal with the problems and challenges faced in developing batteryless implantable biomedical devices.

## 5.2 METHODS

In the literature, various methods are discussed for harvesting energy for implantable biomedical devices. In this chapter, we will discuss all the methods in which the human body is used as a source for scavenging electrical energy to power implantable biomedical devices.

Humans are themselves a significant source of generating the energy required for powering implantable biomedical devices. For example, approximately 81 mW of power can be harvested even while a human is sleeping [4]. Figure 5.1 shows various energies produced through the human body, which can be used for harvesting energy for implantable biomedical devices.

### (A) Kinetic Energy

The human body can easily generate kinetic energy while it is in motion. For example, when a person is walking or running, kinetic energy is built up. This energy starts from their feet. There are various batteryless devices that use this kinetic energy generated through human movement. For instance, npower personal energy backup chargers draw their power from kinetic energy through human body movement (walking and running). To convert kinetic energy into electrical energy to power implantable biomedical devices, different transducers are used. These are grouped into three categories: piezoelectric effect, electrostatic induction and electromagnetic induction [3, 5].

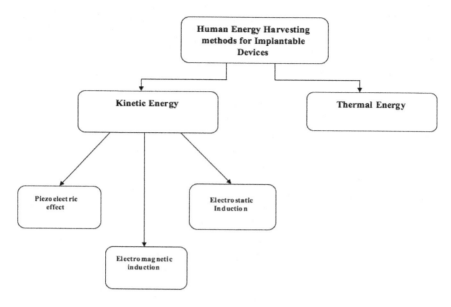

FIGURE 5.1  Human body energy-harvesting methods used in implantable biomedical devices

## 5.2.1 Piezoelectric Effect

The piezoelectric effect is the property of some materials to generate electrical energy through mechanical motion. When these materials are placed under mechanical stress, shifting of the charges inside the material takes place, which leads to the generation of electrical energy.

The brothers Pierre and Jacques Curie first discovered the piezoelectric effect in 1880 by combining their knowledge of pyroelectricity with crystal structure. During World War I, the first application of the piezoelectric effect was developed: sound navigation and ranging (SONAR).

For piezoelectric biomedical devices, the main concern is the selection of piezoelectric material. The selection of material is made mostly on two factors: the strength of the piezoelectric effect generated by material and the cost. In case of implantable piezoelectric biomedical devices, this selection process faces some other constraints as well, like flexibility of material due to direct human contact. Biocompatible piezoelectric materials are used for these devices, as they should not contain toxic metals like lead [6]. The most common biocompatible piezoelectric materials are ceramic (quartz, titanium) and polyvinylidene fluoride (PVDF) copolymers. Table 5.1 lists some of the studies done on piezoelectric implantable biomedical devices.

TABLE 5.1    Literature on Piezoelectric Implantable Biomedical Devices

| Year | Authors | Summary of Work Done | Refs. |
|------|---------|----------------------|-------|
| 1998 | Paradiso et al. | To generate power of 1 W, they implement a spring magnetic generator in a shoe heel. This prototype generates a good amount of power but comes with a limitation: it is applicable only for patients who can walk | [7] |
| 2001 | Shenck and Paradiso | They further work on their last prototype and integrate piezoelectric elements in two positions this time: first on the heel and second on the toe of the shoe. This model generates 8.3 mW from the heel and 1.3 mW from the toe. The limitation of this device remains the same | [8] |
| 2002 | Kornbluh et al. | To improve the results obtained from Paradiso et al., they further compress the piezoelectric material. In this prototype, it is placed in the heel of the shoes. From this prototype, the amount of power generated increases, but the same problem continues | [9] |
| 2001 | Ramsay and Clark | They implement a square PZT-5A piezoelectric material to convert the mechanical strain into electrical energy. They use fluctuation in blood pressure as a mechanical strain applied to piezoelectric material. Through this device, they are able to generate a maximum power of 2.3 μW | [10] |
| 2005 | Sohn et al. | They also use blood pressure to generate electric energy through piezoelectric materials. Here they implement circular and square polyvinylidene fluoride plates to generate energy | [11] |
| 2005 | Platt et al. | They embed piezoelectric ceramics within orthopedic implants to generate energy. They are able to produce a power of 4.8 mW | [12] |
| 2007 | Hong et al. | They also do work similar to that done by Platt et al. They embed piezoelectric ceramics within knee replacement implants to generate energy. They are able to produce power of 1.2 mW | [13] |
| 2010 | Shaban et al. | They improve the work done by Hong et al. To make the improvement, they embed four piezoelectric ceramics within knee replacement implants to generate energy. They are able to produce power of 1.81 mW | [14] |
| 2015 | Lu et al. | In this research, the authors give a prototype of energy harvesting from the heart. They design a device made of piezoelectric material zirconate titanate (PZT) ceramic. For harvesting energy from piezoelectric material, the motion of the heart is used | [15] |

| Year | Authors | Summary of Work Done | Refs. |
|------|---------|---------------------|-------|
| 2018 | Dong et al. | The authors propose a batteryless pacemaker. They design a pacemaker draped with flexible porous polyvinylidene fluoride-trifluoroethylene thin film within a dual-cantilever structure. This structure consists of two free ends projecting out for scavenging energy through heart motion. The maximum power generated is 0.5 V and 43 nA under a frequency of 1 Hz | [16] |
| 2020 | Xu et al. | They design self-sustainable pacemakers. The pacemakers harvest energy through the heartbeat using multibeam piezoelectric composite thin films. Here they employ polydimethylsiloxane (PDMS)-infilled microporous P (VDF-TrFE) composite films | [17] |

## 5.2.2 Electrostatic Induction

To produce electrical energy for biomedical implantable devices through mechanical motion, electrostatic generators are used. These generators work on the principle of electrostatic induction. Generally, electrostatic generators consist of two conducting plates, also known as electrodes, which are isolated electrically through air, vacuum or dielectric insulators to form a capacitor-like structure. In this structure, one plate is made fixed and the other one is movable. The distance between these two electrodes changes through the movement of movable electrodes via the motion of the human body. The change in distance results in the generation of electrical charge [18]. Figure 5.2 shows a block diagram of commonly used electrostatic generators for powering implantable medical devices (IMDs).

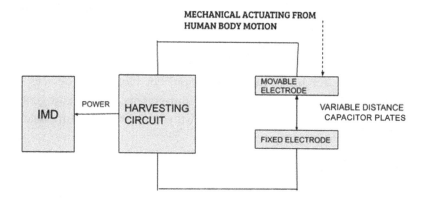

FIGURE 5.2    Block diagram of electrostatic generator

TABLE 5.2    Literature on Electrostatic Generators

| Year | Authors | Summary of Work Done | Refs. |
|------|---------|----------------------|-------|
| 2002 | Tashiro et al. | An electrostatic generator is proposed by the authors in this study, which generates electrical energy through the motion of a movable plate through the force generated by cardiac signal. In this study, they employ a MEMS capacitor with variable capacitors ranging from 32 to 200 nF. Approximately 58 μW of power is produced by this generator | [20] |
| 2001 | Meninger et al. | In this study, the authors are able to produce power of 8 μW by using a micro-machined capacitor (MMC) in their electrostatic generator. In addition to this capacitor, to improve the function, a parallel capacitor is also employed with MMC | [21] |
| 2006 | Miao et al. | The authors propose an electrostatic generator, which produces power of approximately 80 μW. They use non-resonant MEMS in the capacitor structure | [22] |
| 2009 | Elfrink et al. | The authors propose an electrostatic generator, which produces power of approximately 60 μW. They use aluminum nitride with an unpackaged device to provide capacitor structure | [23] |
| 2021 | Rodríguez González | The author, in her final degree project, works on manufacturing a U-TENG as a power source for medical devices. It demonstrates the triboelectric effect, as voltage appears when applying a periodic force to change separation between plates. However, the maximum open circuit voltage obtained is much lower compared with the ones described in the article, 5.3 V compared to 480 V | [24] |

Electrostatic generators work under two operating modes, constant charge mode and constant voltage mode [19]. In constant charge mode, movement of the movable plate changes the voltage across the capacitor, and in constant voltage mode, this movement changes the charge across the capacitors. One of the major drawbacks of these generators is they are not suitable for high power generation. For biomedical devices like implantable biosensors with low power requirements, electrostatic generators work well for this reason. Table 5.2 lists some of the literature on electrostatic generators.

## 5.2.3 Electromagnetic Induction
To power biomedical implantable devices, electromagnetic generators are also employed. Electromagnetic generators work on the principle of

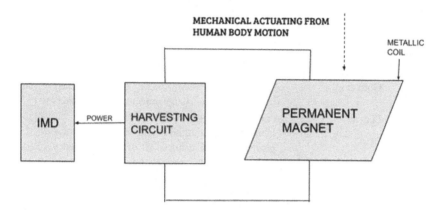

FIGURE 5.3   Block diagram for electromagnetic generator

electromagnetic induction based on Faraday's law [25]. The general construction of electromagnetic generators consists of a permanent magnet and a coil. The relative motion between a coil and permanent magnet results in the generation of magnetic flux. The rate of change of this magnetic flux produces electromagnetic force (EMF). To create the relative motion between a coil and magnet in electromagnetic generators, human body motion is exploited, as shown in Figure 5.3.

Electromagnetic generators work in two ways. In the first, relative motion is utilized, while the generator system remains immobile, and in the second, rigid body motion is employed with the internal force of a weight on the ground [26].

### (B) Thermal Energy

Thermoelectric generators translate thermal energy into electrical energy using the Seebeck effect [3]. This forms the principle for thermocoupling. The energy harvested through thermoelectric generators is enough to power some implantable biomedical devices like muscle stimulators, cochlear hearing replacements and so on. The Seebeck effect utilizes the temperature difference between two electrical conductors or semiconductors to generate power. Using this principle, the general construction of thermoelectric generators consists of a large number of thermocouples connected electrically in series and thermally in parallel to form thermopiles. This construction of thermoelectric generators is ideal for scavenging energy from the temperature difference generated within the human body. Here, the energy produced through the generator is limited by the

TABLE 5.3    Studies on Electromagnetic Generators

| Year | Authors | Summary of Work Done | Refs. |
|---|---|---|---|
| 1998 | Amirtharajah and Chandrakasan | Use electromagnetic VDRG built to generate 400 μW of power | [27] |
| 1999 | Goto et al. | Authors use the "Seiko Kinetic" approach, which uses heartbeats to charge implantable pacemakers | [28] |
| 2000 | Li et al. | A new electromagnetic MEMS VDRG is fabricated by the authors to generate 10 μW of power at 2 V DC using 64 Hz of input frequency | [29] |
| 2001 | Williams et al. | The authors use the same prototype given in [30] to generate 0.3 μW of power from a 4-MHz excitation input | [31] |
| 2009 | Romero et al. | The authors use an axial flux generator to generate energy by electromagnetic indication. This generator involves a gear-shaped, planar coil and a ring-attached eccentric weight. This device, based on a rigid body, is fixed on the ankle and, during walking, is used to provide 3.9 μW of power to small biomedical devices | [32] |
| 2011 | Nasiri et al. | The abdomen moves with a frequency of 0.3 Hz during breathing, producing power of about 1.1 mW through an electromagnetic generator | [33] |
| 2018 | Tholl et al. | Develop a novel generator that converts electrocardial heart motion into electrical energy by electromagnetic induction | [30] |
| 2018 | Angelika-Nikita et al. | Design a electromagnetic induction device which converts arterial wall pulsation due to blood flow into electrical energy | [34] |

TABLE 5.4    Studies on Electromagnetic Generators

| Year | Authors | Summary of Work Done | Refs. |
|---|---|---|---|
| 1999 | Stark and Stordeur | Design a thermopile using human body heat to generate power of approximately 1.5 μW when a thermal difference of 5K is exploited | [35] |
| 2004 | Strasser et al. | They perform work similar to that conducted by Stark and Stordeur and produce power of approximately 1 μW | [36] |
| 2018 | Tupe et al. | Develop a thermoelectric generator using polydimethylsiloxane substrate and thermocouples, working as a power supply system for implantable biomedical devices. It is able to produce power of 50 nW when the temperature difference between the ambient environment and body temperature is 7°C | [37] |

Carnot efficiency [$\eta = (Th - Tc)/Th$] where $Th$ – Carnot efficiency defines the upper limit on the efficiency of thermoelectric devices while converting thermal energy into electrical energy.

The literature discusses various thermoelectric energy-harvesting generators using human body heat to generate power. However, these generators come with a limitation of generating only a few hundred microwatts of power when exploited to temperature gradients below 5K.

## 5.3 CONCLUSION

In this chapter, numerous techniques were discussed for harvesting energy for implantable biomedical devices. The environment and humans themselves are the two main sources of harvesting energy for implantable biomedical devices. All methods related to human sources used to date for energy harvesting in biomedical implantable devices are thoroughly reviewed in this chapter. Some issues and challenges confronted while developing batteryless implantable biomedical devices were also discussed. In a nutshell, this chapter has provided readers a history to deal with the troubles and challenges faced when developing batteryless implantable biomedical devices.

## REFERENCES

[1] M. Billinghurst and T. Starner, "Wearable devices. New ways to manage information," *IEEE Journals and Magazines (Computer)*, vol. 329, no. 1, pp. 57–64, 1999.

[2] M. E. Kiziroglou and E. M. Yeatman, "Materials and techniques for energy harvesting," in *Functional Materials for Sustainable Energy Applications*, Imperial College London, pp. 541–572, 2012.

[3] M. A. Hannan, S. Mutashar, S. A. Samad and A. Hussain, "Energy harvesting for the implantable biomedical devices: Issues and challenges," *Biomedical Engineering Online*, vol. 13, no. 1, pp. 1–23, 2014.

[4] J. Paulo and P. D. Gaspar, "Review and future trend of energy harvesting methods for portable medical devices," in *World Congress on Engineering (WCE)*, pp. 1–6, London, UK, 2010.

[5] E. Romero, R. O. Warrington and M. R. Neuman, "Energy scavenging sources for biomedical sensors," *Physiological Measurement*, vol. 30, no. 9, pp. 35–62, 2009.

[6] A. H. Rajabi, M. Jaffe and T. L. Arinzeh, "Piezoelectric materials for tissue regeneration: A review," *Acta Biomaterialia*, vol. 24, pp. 12–23, 2015.

[7] J. Paradiso, J. Kymissis, C. Kendall and N. Gershenfeld, "Parasitic power harvesting in shoes," *IEEE International Symposium on Wearable Computers*, vol. 24, pp. 132–139, 1998.

[8] N. S. Shenck and J. A. Paradiso, "Energy scavenging with shoe-mounted piezoelectric," *IEEE Micro*, vol. 21, no. 3, pp. 30–42, 2001.

[9] R. D. Kornbluh, R. Pelrine, Q. Pei, R. Heydt, S. Stanford, S. Oh and J. Eckerle, "Electroelastomers: Applications of dielectric elastomer transducers for actuation, generation, and smart structures," *Smart Structures and Materials (SPIE)*, pp. 254–270, San Diego, 2002.

[10] M. J. Ramsay and W. W. Clark, "Piezoelectric energy harvesting for bio-MEMS applications," in *Proceeding of the Smart Structures and Materials (SPIE)*, pp. 429–438, Newport Beach, CA, 2001.

[11] J. W. Sohn, S. B. Choi and D. Y. Lee, "An investigation on piezoelectric energy harvesting for MEMS power sources," *Journal of Mechanical Engineering Science*, vol. 219, no. 4, pp. 429–436, 2005.

[12] S. R. Platt, S. Farritor, K. Garvin and H. Haider, "The use of piezoelectric ceramics for electric power generation within orthopedic implants," *IEEE ASME Trans Mechatron*, vol. 10, no. 4, pp. 455–461, 2005.

[13] C. Hong, J. Chen, Z. Chun, W. Zhihua and L. Chunsheng, "Power harvesting with PZT ceramics," in *The International IEEE Symposium on Circuits and Systems (ISCAS)*, pp. 557–560, Korea, 2007.

[14] A. Shaban, G. Manuel, H. Chafiaa, S. Eric, and R. Christian, "Self-powered instrumented knee implant for early detection of postoperative complications," in *Proceeding of the 32nd Annual International Conference of the IEEE (EMBS)*, pp. 121–5124, Buenos Aires, Argentina, 2015.

[15] B. Lu, Y. Chen, D. Ou, H. Chen, L. Diao, W. Zhang, J. Zheng, W. Ma, L. Sun and X. Feng, "Ultra-flexible piezoelectric devices integrated with heart to harvest the biomechanical energy," *Scientific Reports*, vol. 5, no. 1, pp. 1–9, 2015.

[16] L. Dong, X. Han, Z. Xu, A. B. Closson, Y. Liu, C. Wen, X. Liu, G. P. Escobar, M. Oglesby, M. Feldman and Z. Chen, "Flexible porous piezoelectric cantilever on a pacemaker lead for compact energy harvesting," *Advanced Materials Technologies*, vol. 4, no. 1, January 2019.

[17] Z. Xu, C. Jin, A. Cabe, D. Escobedo, N. Hao, I. Trase, A. B. Closson, L. Dong, Y. Nie, J. Elliott and M. D. Feldman, "Flexible energy harvester on a pacemaker lead using multibeam piezoelectric composite thin films," *ACS Applied Materials & Interfaces*, vol. 12, no. 30, pp. 34170–34179, 2020.

[18] M. Miyazaki, H. Tanaka, T. Nagano, N. Ohkubo and T. Kawahara, "Electric-energy generation through variable-capacitive resonator for power-free LSI," *IEICE Trans. Electron*, vol. 87, pp. 549–555, 2004.

[19] N. C. Tsai and C. Y. Sue, "Review of MEMS-based drug delivery and dosing systems," *Sensors and Actuators A: Physical*, vol. 134, pp. 555–564, 2007.

[20] R. N. Tashiro, K. Kabei, Y. Katayama, F. Ishizuka, K. Tsuboi and B. Tsuchiya, "Development of an electrostatic generator that harnesses the ventricular wall motion," *International Journal of Artificial Organs*, vol. 5, pp. 239–245, 2002.

[21] S. Meninger, J. Mur-Miranda, R. Amirtharajah, A. P. Chandrasakan and J. H. Lang, "Vibration to electric energy conversion," *IEEE Trans Very Large Scale Integration (VLSI) System*, vol. 9, no. 1, pp. 48–53, 2001.

[22] P. Miao, P. Mitcheson, A. Holmes, E. Yeatman, T. Green and B. Stark, "MEMS inertial power generators for biomedical applications," *Microsystem Technology*, vol. 12, no. 10–11, pp. 1079–1083, 2006.

[23] R. Elfrink, T. M. Kamel, M. Goedbloed, S. Matova, D. Hohlfeld, V. Y. Andel and V. R. Schaijk, "Vibration energy harvesting with aluminium nitride-based piezoelectric devices," *Journal of Micromechanics and Microengineering*, vol. 19, no. 9, p. 094005, 2009.

[24] H. Rodríguez González, *Study of Systems Powered by Triboelectric Generators for Bioengineering Applications*, M. S Thesis, Dept of Biomedical Engg., University of Barcelona, Spain. 2021.

[25] A. B. Amar, A. B. Kouki and H. Cao, "Power approaches for implantable medical devices," *Sensors*, vol. 15, no. 11, pp. 28889–28914, 2015.

[26] J. Paulo and P. Gaspar, "Review and future trend of energy harvesting methods for portable medical devices," in *Proceedings of the World Congress on Engineering*, pp. 168–196, London, UK, July 2010.

[27] R. Amirtharajah and A. Chandrakasan, "Self-powered signal processing using vibration-based power generation," *IEEE Journal of Solid State Circular*, vol. 33, no. 5, pp. 687–695, 1998.

[28] H. Goto, T. Sugiura, Y. Harada and T. Kazui, "Feasibility of using the automatic generating system for quartz watches as a leadless pacemaker power source," *Medical & Biological Engineering & Computing*, vol. 37, no. 1, pp. 377–380, 1999.

[29] W. Li, T. Ho, G. Chan, P. Leong and H. Y. Wong, "Infrared signal transmission by a laser-micromachined, vibration-induced power generator," in *Proceedings of the 43rd IEEE Midwest Symposium on Circuits and Systems*, pp. 236–239, Lansing, MI, USA, 2000.

[30] M. V. Tholl, A. Haeberlin, B. Meier, S. Shaheen, L. Bereuter, B. Becsek, H. Tanner, T. Niederhauser and A. Zurbuchen, "An intracardiac flow based electromagnetic energy harvesting mechanism for cardiac pacing," *IEEE Transactions on Biomedical Engineering*, vol. 66, no. 2, pp. 530–538, 2018.

[31] C. Williams, C. Shearwood, M. Harradine, P. Mellor, T. Birch and R. Yates, "Development of an electromagnetic micro-generator," *IET Journal of Magazine 2001*, vol. 148, no. 6, pp. 337–342, 2001.

[32] E. Romero, R. O. Warrington and M. R. Neuman, "Body motion for powering biomedical devices," in *Proceeding of the 31st Annual International Conference of the IEEE (EMBS)*, pp. 2752–2755, Minnesota, Minneapolis, 2009.

[33] A. Nasiri, S. A. Zabalawi and D. C. Jeutter, "A linear permanent magnet generator for powering implanted electronic devices," *IEEE Transaction on Power Electron*, vol. 26, pp. 192–199, 2011.

[34] M. Angelika-Nikita, N. Hadjigeorgiou, C. Manopoulos and J. Georgiou, "Converting energy captured from blood flow into usable electric power: Design optimisation," arXiv preprint arXiv, 1809, 10929, 2018.

[35] I. Stark and M. Stordeur, "New micro thermoelectric devices based on bismuth telluride-type thin solid films," in *Proceeding of the 18th International Conference Thermoelectric Baltimore*, pp. 465–472, Baltimore, MD, 1999.

[36] M. Strasser, R. Aigner, C. Lauterbach, T. F. Sturm, M. Franosch and G. Wachutka, "Micromachined CMOS thermoelectric generators as on-chip power supply," *Sensor Actuator A-Phys*, vol. 114, no. 2–3, pp. 362–370, 2004.

[37] S. Tupe, T. Shaikh, D. T. Kashid and D. S. Ghodake, "Study on development of power supply system for medical implants using thermoelectric energy harvesting from human body," *International Journal for Research in Engineering Application & Management (IJREAM)*, Special Issue—AMET-2018, ISSN: 2454–9150.

# CHAPTER 6

# Practices Involved in Broadband Vibration Energy Harvesting Utilizing Variable-Length Multi-Reed Cantilever Array

Rathishchandra R. Gatti

## CONTENTS

DOI: 10.1201/9781003218760-6

## 6.1 INTRODUCTION

Vibration energy harvesters (VEHs) are energy transducers that convert the energy from vibrations into useful electrical energy [1]. Typical energy transduction mechanisms used in VEHs are electrostatic [2], electromagnetic [3], piezoelectric [4] and magnetostrictive [5] energy conversions. Of these, piezoelectric VEHs are popular in harvesting vibration energies in high-frequency low amplitudes, while electromagnetic VEHs are suitable for harnessing low-frequency, high-amplitude vibrations [6].

VEHs are very useful in powering wireless sensors, making them autonomous. A state-of-the-art example is General Electric's Insight Mesh wireless sensor network installed in a Shell power plant in Nevada that is powered by electromagnetic VEHs [7]. Other useful applications of VEHs include powering wireless structural health monitoring [8], wearable devices [9] and in-vivo wireless body area network [10] sensors.

One of the characteristics of ambient vibrations that are available is their randomness, constituting many vibrations of different frequencies superimposed, which are hard to distinguish unless fast Fourier transform (FFT) or other spectral analyses [11] is performed to find the prominent frequencies. VEHs generate maximum power only when the vibration frequency of the source is in resonance with the natural frequency of the VEH. This is not always the case due to the presence of random vibrations. Hence, one of the key requirements in the design of energy harvesters is their ability to charge an appreciable amount of power for a broad range of frequencies. Over the years, a number of structural topologies have been developed in both piezoelectric and electromagnetic VEHs for generating energy for a range of source frequencies. Amongst these, cantilever structures are commonly used to fabricate micro-electro mechanical systems (MEMS)-scale vibration energy transducers, especially piezoelectric-based sensors and energy harvesters.

The earliest patent filed that cited the concept of energy harvesting of devices was in 1973 for a biomedical implant which included an embedded energy transducer that harvests energy from mammalian bodies (EP0761256A3) [12]. In 2000, the Seiko Epson Corporation filed a patent (US 6097675) [13] on meso-scale energy generators embedded in watches that became popular as self-powered watches. Around 2003, the University of Florida came up with the concept of integrated MEMS piezoelectric-based micro-energy harvesters coupled to a power processor, which is capable of stabilizing the output power from variations of input random vibrations (WO/2003/096444A2) [14]. In 2005, the method of using frequency up-conversion to harness low-frequency vibrations was patented (WO/2005/069959A2) by Kulah and Najafi [15]. To capture the vibrational energy along all three directions ($X$, $Y$ and $Z$), Priya and Myers designed a multidirectional VEH (US 20080174273) [16] that consisted of plural piezoelectric cantilevers with bases normal to each of the directions. A patent closer to the proposed idea of a multi-reed VEH was suggested by Malkin and Davis in US 6858970 [17], as depicted in Figure 6.1. The energy harvester consisted of piezoelectric cantilevers of varied lengths but the same excitations resonating at different primary frequencies to their base [18]. Similarly, another patent (US 7239066) [18] consisted of plural piezoelectric devices arranged in a circular manner and actuated by a single actuator. The natural frequency of the cantilever spring or the single-degree-of-freedom system that adopts the cantilever spring depends on its length. A novel design incorporating a multi-reed cantilever array

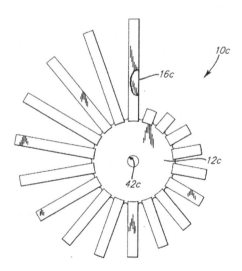

FIGURE 6.1   Multi-cantilever vibration energy harvester

consisting of different lengths of cantilevers with lumped masses arranged in a circular pattern is proposed in this design to capture energies from different frequencies. The array was initially designed and analyzed, and then a prototype was fabricated and tested for validation.

## 6.2 DESIGN

The proposed design consists of an array of eight cantilevers of different lengths arranged in a circular pattern forming equal angles of 45° to each other with a common fixed base, as shown in Figure 6.2. Except the effective length of the cantilever ($l_e$), all the other parameters such as width and thickness of the cantilever beam and mass of the suspended masses are kept equal for all eight cantilevers. The effective lengths of the eight cantilevers start from 10 to 80 mm in steps of 10 mm. The effective length is considered from the edge of the fixed support to the middle plane of the lumped mass.

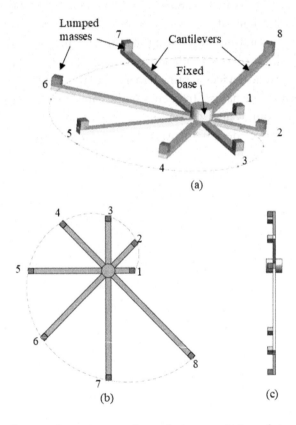

FIGURE 6.2  Proposed energy transducer design consisting of circular array of variable-length cantilevers (a) isometric view, (b) top view, and (c) side view

TABLE 6.1    Design Parameters of the Variable-Length Cantilever Arrays

| | |
|---|---|
| **Lumped Mass Specifications** | |
| Lumped mass material | Plain carbon steel |
| Mass of the lumped mass | 0.47 g |
| **Cantilever and Base Specifications** | |
| Cantilever and fixed base material | Polyvinylidene difluoride |
| Elastic modulus | 2,450 N/mm² |
| Poisson's ratio | 0.18 |
| Shear modulus | 318.9 N/mm² |
| Tensile strength | 52 N/mm² |
| Mass density ($\rho$) of PVDF | 1,770 kg/m³ |
| Width of the cantilever ($w$) | 4 mm |
| Thickness of the cantilevers ($t$) | 2 mm |
| Area moment of inertia ($I = wt^3/12$) | 2.667e-12 mm⁴ |
| Effective lengths of the eight cantilevers ($l_e$) | (10,20,...,70,80) mm |
| Masses of the cantilevers ($m_e$) | {0.198,0.340,0.481,0.623,0.765,0.906,1.000,1.200} g |

The natural frequency of the eight cantilevers was derived as

$$f_n = \frac{1}{2\pi}\sqrt{\frac{3EI}{(m_i + m_c/3)l_e^3}},$$

$$(1)$$

where $E$ = modulus of elasticity of the cantilever material, $I$ = area moment of inertia of the cantilever, $m_l$ = mass of the lumped mass, $m_c$ = mass of the cantilever spring and $l_e$ = effective length of the cantilever. For eight cantilevers, only the factors $m_c$ and $l_e$ vary with the change in length $l_e$. The different values of the design parameters are summarized in Table 6.1.

## 6.3 NUMERICAL SIMULATION

Two types of simulation were performed—analytical simulation on the equations of motion using MATLAB to find the fundamental natural frequencies and numerical finite-element analysis (FEA) simulation using the Solidworks linear dynamics simulation package to cross-validate the modal analysis results. Once the natural frequencies of the eight cantilevers were validated, further investigations on the Von-Mises stresses and the maximum displacement were analyzed.

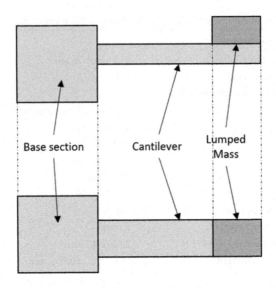

FIGURE 6.3   Piecewise FEA simulation model

## 6.3.1  Piecewise FEA Simulation Model

The design of circular array of variable-length cantilevers is divided into eight parts to simplify the model for FEA simulation. Each part is as shown in Figure 6.3. The simplified model has a fixed base as 1/8th of a segment of the cylindrical fixed base. The effective length of the cantilever progressively increases from 10 to 80 mm for eight study iterations.

## 6.3.2  FEA Setup and Assumptions

Linear dynamic analysis was performed using modal time history analysis of five modes considering higher modes impractical for seismic energy transducers. Rayleigh damping was considered where the damping matrix (**D**) is given by (**D**) = $\alpha(M) + \beta(K)$, where $\alpha$ and $\beta$ were assumed to be 0.02 and 0.04, respectively. A lab temperature of 25°C was assumed during the simulation. The fixed base sector surface as shown in Figure 6.3 was fixed using a fixed geometry constraint. A base excitation load of 2 mm amplitude was given to the fixed base sector surface assuming the average value of 0–4 mm amplitude vibrations typically found in most rotating machinery.

The simplified model was meshed using the solid curvature-based tetrahedral element with four Jacobian points (sufficient for a straight model) and aspect ratio of less than 3 for 99.8% of the elements and maximum aspect ratio of 3.5 to 4.5 for the change in length. The meshed model contained around 18,061 nodes and 11,096 elements (for a 50-mm cantilever, the number of nodes and elements vary when the length is iterated for 8 cantilevers). The average element size was 0.74 mm.

### 6.3.3 FEA Solver and Method of Running the Analysis

Although approximate compared to the direct sparse solver, the iterative FFE plus solver was used because of time and memory constraints. The dynamic analysis using a random vibration of 1 mm/Hz at base excitation was performed in the time range from 0 to 1 s with an increment of 0.01 s. Eight different studies were conducted for eight cantilevers by varying the length of the cantilever of the simplified FEA model for the same initial FEA assumptions as discussed, with re-meshing and rerunning the FEA dynamic analysis for each study.

## 6.4 EXPERIMENTATION

The experimentation was performed using a meso-scale prototype made of cantilevers of polyvinylidene fluoride (PVDF) material that meets ASTM 6713 standards with glued plain carbon steel cubes, as shown in Figure 6.2. The prototype was mounted on a vibration exciter with an integrated power amplifier and was excited using a function generator, as shown in the schematic Figure 6.4. For each trial, two accelerometers were fixed, one at the fixed base for measuring the base excitation and the other at the steel lump mass of each cantilever for measuring the cantilever excitation.

### 6.4.1 Experimental Setup

The experimental setup is as shown in Figure 6.4. The prototype is initially mounted on a integrated vibration exciter-power amplifier, which is connected to a function generator for base excitation of desired frequency

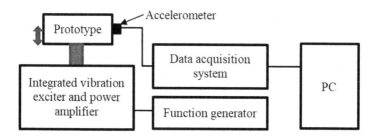

FIGURE 6.4   Schematic of the experimental setup

and amplitude. The uni-axial accelerometer is connected to each of the cantilevers according to the experimental set. The data is read by a data acquisition system with a vibration module and is connected to the lab computing system (PC) for data acquisition and visualization.

### 6.4.2 Experiment

There were eight sets of trials for eight cantilevers, and each cantilever was excited with the amplitude of 0.1 to 1 mm, where the amplitudes were selected based on the cantilever length to avoid violent vibrations during resonance. This method was adopted since the amplitude of vibration does not determine the resonance. The prototype was excited using the vibration exciter from 5 to 100 Hz in steps of 5 Hz and then in steps of 1 Hz at the resonant frequency range as predicted by numerical values to determine the resonant frequency at a resolution of 1 Hz.

## 6.5 RESULTS AND DISCUSSION

Since the main interest of the energy transducer is its resonant behavior with a range of source frequencies provided by random vibrations, modal time history analysis was initially done to find the modal frequencies of the eight cantilevers. This was followed by Von-Mises stress analyses and the maximum displacements of the eight cantilevers.

### 6.5.1 Comparison of the Analytical, Numerical and Experimental Values of Fundamental Frequencies of Different Cantilevers

First, the theoretically obtained natural frequencies of the cantilevers at their fundamental modes of vibration were compared with the experimentally determined natural frequencies, as shown in Table 6.2.

It can be observed that the natural frequency values are in agreement from Cantilever 3 ($l_e$ = 30 mm) to Cantilever 8($l_e$ = 80 mm). This is true since the approximation in the analytical equation (Equation 1) can be considered accurate for longer lengths of the cantilever. The aspect ratio of the effective length of the cantilever to the length of the fixed base is smaller compared to the lengths of Cantilever 3 and above. It can be observed that the radius of the fixed base is 8 mm and the effective length is 20 mm where the analytical and numerical values do not agree. Thus, it can be empirically concluded that the proposed design is validated for an aspect ratio of fixed support diameter to cantilever effective length of 30/8 = 3.75 and above. It was observed that the fundamental mode frequencies were

TABLE 6.2   Comparison of the Numerical and Analytical Fundamental Natural Frequencies

| Cantilever number | Effective Length, $l_e$ (mm) | Natural Frequency at Fundamental Mode | | |
|---|---|---|---|---|
| | | $f_n$ (Analytical) (Hz) | $f_n$ (Numerical) (Hz) | $f_n$ (Experimental) (Hz) |
| 1 | 10 | 962.35 | 760.35 | Not tested |
| 2 | 20 | 326.18 | 300.51 | Not tested |
| 3 | 30 | 170.77 | 165.25 | Not tested |
| 4 | 40 | 106.99 | 106.25 | 102 |
| 5 | 50 | 74.022 | 74.77 | 71 |
| 6 | 60 | 54.56 | 55.82 | 50 |
| 7 | 70 | 42.03 | 43.44 | 40 |
| 8 | 80 | 33.45 | 34.87 | 31 |

almost linear from Cantilevers 4 to 8, whose lengths varied from 40 to 80 mm. This suggests that the energy harvester can be designed after knowing the typical range of prominent frequencies obtained from the FFT of the random source vibrations. If the source vibrations fall between two values, say, $f_1$ and $100 f_1$, then a range of $n$ cantilevers can be designed with lengths varying between $y_1$ and $y_2$, provided the range ($f_1$, $y_1$) to ($100 f_1$, $y_2$) is sufficiently linear, as observed in Table 6.2 between (50, 74.77) and (80, 34.87).

## 6.5.2  Modal Frequencies of Higher Modes of Different Cantilevers

An energy transducer is typically designed to operate at fundamental mode. However, the knowledge of higher modes will be useful to observe if the values of any frequencies of the lower mode are in proximity with the higher-mode frequencies, as seen in Figure 6.5. For example, the first mode frequency (298 Hz) of Cantilever 2 ($l_e$ =20 mm) is in proximity with the second mode frequency (300 Hz) of the Cantilever 3 ($l_e$ = 30 mm). Therefore, if a multi-reed energy harvester with these two lengths of cantilever vibrates at a source frequency of ~300 Hz, then Cantilever 2 vibrates at the first mode and Cantilever 3 at the second mode. This needs to be considered in a multi-reed energy harvester design, especially if it is possible to generate power in the second mode of vibration. Similar proximity pairs can be identified as shown in Figure 6.5(a).

It can be observed in Figure 6.5(b) that the frequencies of higher modes for short cantilevers is very high and unsafe for the structural integrity of the cantilever design. It can also be observed that longer cantilevers from

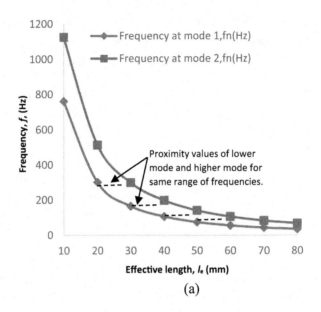

FIGURE 6.5 (a) Fundamental and second-mode frequencies

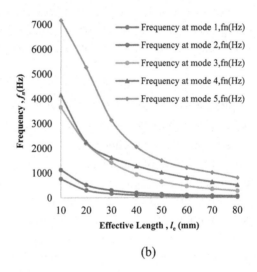

FIGURE 6.5 (b) Higher-mode frequencies of the eight cantilevers

Cantilever 4 ($l_e$ = 4) to Cantilever 8 ($l_e$ = 8) have lower frequencies. Hence, longer cantilevers with a high aspect ratio of effective length/fixed base length are structurally safe for energy harvester design. Another advantage of the longer cantilevers is the existence of more proximity points, making the design achieve maximum energies for a broad range of frequencies.

### 6.5.3 Stresses and Displacements of Different Cantilevers

Von-Mises stresses were considered for analyzing the structural integrity of the cantilevers. As evident from Figure 6.6, the stresses are maximum at the intersection of the fixed edges of the cantilevers and the fixed base. Stresses throughout the length of the cantilevers decrease with the increase in the effective length, as observed in Cantilever 4 to Cantilever 8.

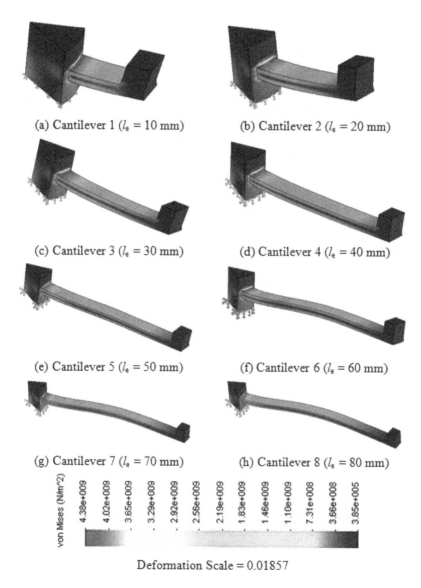

(a) Cantilever 1 ($l_e$ = 10 mm)     (b) Cantilever 2 ($l_e$ = 20 mm)

(c) Cantilever 3 ($l_e$ = 30 mm)     (d) Cantilever 4 ($l_e$ = 40 mm)

(e) Cantilever 5 ($l_e$ = 50 mm)     (f) Cantilever 6 ($l_e$ = 60 mm)

(g) Cantilever 7 ($l_e$ = 70 mm)     (h) Cantilever 8 ($l_e$ = 80 mm)

von Mises (N/m^2)

4.38e+009  4.02e+009  3.65e+009  3.29e+009  2.92e+009  2.56e+009  2.19e+009  1.83e+009  1.46e+009  1.10e+009  7.31e+008  3.66e+008  3.85e+005

Deformation Scale = 0.01857

FIGURE 6.6    Von-Mises stresses of the eight cantilevers

TABLE 6.3   Maximum Displacement and Von-Mises Stresses of the Eight Cantilevers

| Cantilever Number | Effective Length (mm) | Maximum Displacement (mm) | Maximum Stress (N/mm²) |
|---|---|---|---|
| 1 | 10 | 3.60 | 14.85 |
| 2 | 20 | 8.38 | 12.60 |
| 3 | 30 | 9.44 | 7.71 |
| 4 | 40 | 9.66 | 7.00 |
| 5 | 50 | 9.52 | 7.88 |
| 6 | 60 | 9.12 | 8.19 |
| 7 | 70 | 8.74 | 7.72 |
| 8 | 80 | 9.00 | 8.14 |

The post-processed data of maximum Von-Mises stresses and the maximum displacements of the eight cantilevers are summarized in Table 6.2. The maximum Von-Mises stresses decreased from 14.82 to 8.14 N/mm² from Cantilever 1 ($l_e = 10$ mm) to Cantilever 8 ($l_e = 80$ mm). This is in agreement with the observation made earlier in Figure 6.6. The maximum displacement increased from 3.60 mm for Cantilever 1 ($l_e = 10$ mm) to 9.00 mm for Cantilever 8 ($l_e = 80$ mm). The maximum displacements to the base excitation ratio ($Z/Y$) were thus 3.60/2 = 1.80 for Cantilever 1 to 9.00/2 = 4.50 for Cantilever 8.

It can also be observed that both the maximum displacement and Von-Mises stress values followed the increasing and decreasing wavering trend with the increase in length of the cantilever. While designing the multi-reed energy harvester, the maximum displacement can be set to a predetermined design value of, say, 5 mm by a combination altering the width of the cantilever beam and effective length of the beam and still maintain the required broadband frequency range.

From this study, it can be observed that increasing the length of the cantilevers in the cantilever array is beneficial to harvest low-frequency vibrations, but it is also restricted by buckling phenomenon and an important design parameter: the power density of the energy harvester. Power density is power generated by the energy harvester to the volume of the space it occupies. A cantilever of longer length will thus decrease the power density. Hence, an optimum length should be designed that is based on the prominent frequencies of the FFTs conducted on random source vibrations as well as the power density.

## 6.6 CONCLUSION

An energy harvester consisting of multiple cantilevers of varying lengths was proposed. The design was initially validated by comparing the analytical, numerical and experimental data of fundamental mode frequencies. Further investigation of the validated model on the Von-Mises stresses and displacements of the eight cantilevers revealed that longer cantilevers can decrease stress and increase displacement. The increase in length is limited by power density and buckling restrictions. This work is very useful as a foundation work for the design of piezoelectric or electromagnetic vibration energy harvesters based on multi-reed cantilevers. Future work can be done on optimizing this design for a specific application such as body energy harvesters for wearables by understanding the prominent frequencies of the source vibrations.

## 6.7 FUTURE SCOPE

Cantilever piezoelectric structures continue to be a favorable choice for developing piezoelectric-based energy harvesters. Only when the yield strength of the cantilever decreases due to weak mechanical properties of the material are piezoelectric beam designs preferred. In the future, a plurality of cantilevers of varied sizes, masses or cross-sections can be explored to achieve broadband energy harvesting from random vibrations. Another area of commercial interest to be explored is the harvesting of energy from low-frequency ambient vibrations where methods such as frequency upconversion can be explored either by mechanical or electrical means.

## 6.8 ACKNOWLEDGMENTS

The author would like to acknowledge the support of Karnataka Government Research Centre, Sahyadri College of Engineering & Management, Mangalore, for carrying out the research on multi-reed cantilever energy harvesters and the Indian Nano User Program, Indian Institute of Technology, Mumbai, for the knowledge gained on PVDF cantilever structures that has led to the design and validation of this new project.

## REFERENCES

[1] R. R. Gatti, "Spatially-varying multi-degree-of-freedom electromagnetic energy harvesting," Doctoral dissertation, Curtin University, Perth, Australia, 2013. http://hdl.handle.net/20.500.11937/161.

[2] P. Basset, D. Galayko, F. Cottone, R. Guillemet, E. Blokhina, F. Marty and T. Bourouina, "Electrostatic vibration energy harvester with combined effect of electrical nonlinearities and mechanical impact," *Journal of Micromechanics and Microengineering*, vol. 24, no. 3, p. 035001, 2014.

[3] R. R. Gatti and I. M. Howard, "Electromagnetic energy harvesting by spatially varying the magnetic field," in *Proceedings of the 2011 2nd International Congress on Computer Applications and Computational Science*, pp. 403–409, Bali, Indonesia, 2011.

[4] A. Nechibvute, A. Chawanda, P. Luhanga and A. R. Akande, "Piezoelectric energy harvesting using synchronized switching techniques," *International Journal of Engineering and Technology*, vol. 2, no. 6, pp. 936–946, 2012.

[5] S. Mohammadi and A. Esfandiari, "Magnetostrictive vibration energy harvesting using strain energy method," *Energy*, vol. 81, pp. 519–525, 2015.

[6] S. Priya and D. J. Inman (Eds.), "Electromagnetic energy harvesting," *Energy Harvesting Technologies*, vol. 21, p. 129, 2009.

[7] L. Hou and N. W. Bergmann, "Induction motor condition monitoring using industrial wireless sensor networks," in *2010 Sixth IEEE International Conference on Intelligent Sensors, Sensor Networks and Information Processing*, pp. 49–54, Brisbane, Australia, 2010.

[8] J. P. Lynch and K. J. Loh, "A summary review of wireless sensors and sensor networks for structural health monitoring," *Shock and Vibration Digest*, vol. 38, no. 2, pp. 91–130, 2006.

[9] W. Wu, S. Bai, M. Yuan, Y. Qin, Z. L. Wang and T. Jing, "Lead zirconate titanate nanowire textile nanogenerator for wearable energy-harvesting and self-powered devices," *ACS Nano*, vol. 6, no. 7, pp. 6231–6235, 2012.

[10] M. S. Taha, M. S. M. Rahim, M. M. Hashim and F. A. Johi, "Wireless body area network revisited," *International Journal of Engineering & Technology*, vol. 7, no. 4, pp. 3494–3504, 2018.

[11] F. Al-Badour, M. Sunar and L. Cheded, "Vibration analysis of rotating machinery using time-frequency analysis and wavelet techniques," *Mechanical Systems and Signal Processing*, vol. 25, no. 6, pp. 2083–2101, 2011.

[12] J. C. Flaherty, "Power supply for implantable device," U.S. Patent 5,810,015, issued 22 September 1998.

[13] O. Takahashi, H. Tatsuo, M. Masatoshi and N. Eiichi, "Electronically controlled mechanical timepiece," U.S. Patent 6,097,675, issued 1 August 2000.

[14] T. Nishida, L. N. Cattafesta III, M. Sheplak and K. D. Ngo, "Resonant energy MEMS array and system including dynamically modifiable power processor," U.S. Patent 6,954,025, issued 11 October 2005.

[15] H. Kulah and K. Najafi, "Method and micro power generator for generating electrical power from low frequency vibrational energy," U.S. Patent 7,579,757, issued 25 August 2009.

[16] S. Priya and R. D. Myers, "Piezoelectric energy harvester," U.S. Patent 7,649,305, issued 19 January 2010.

[17] M. C. Malkin and C. L. Davis, "Multi-frequency piezoelectric energy harvester," U.S. Patent 6,858,970, issued 22 February 2005.

[18] W. F. Ott and T. Edward, "Piezoelectric generators and methods of operating same," U.S. Patent 7,239,066, issued 3 July 2007.

# CHAPTER 7

# Impact of Humidity-Sensing Technology on Clean Energy Generation

*An Overview*

Arshi Salamat, Tarikul Islam

## CONTENTS

## 7.1 INTRODUCTION

In view of rising global concerns about the present situation of planet Earth due to depletion of non-replenishable sources, the advancement in technology of alternative energy generation has become a trending topic among scientists and researchers globally. There are many techniques to generate clean energy, water splitting, solar energy, wind energy, and so on. Inexhaustible sources of energy like wind and solar are friendly to the environment and clean and can be considered potential resources for renewable energy. Many researchers have found that relative air humidity of more than 60% is common in most countries. Therefore, humidity measurement is an important aspect that plays a crucial role in the generation of clean energy, in addition to applications in food processing, pharmaceutical, meteorology, agriculture, home ventilation, and air-conditioning. Humidity influences solar and wind power generation owing to their continuous exposure to open air. Humidity affects the performance of PV cells and wind turbines, which reduces the efficiency of the cell, resulting in a decrease in electrical productivity.

### 7.1.1 Effect of Humidity on Solar Panels

Photovoltaic (PV) cells are greatly influenced by climatic conditions such as radiation from the sun, intensity of radiation, temperature, darkness, dust, dirt, and humidity. Researchers have studied each parameter separately and found solutions to reduce these effects and increase productivity. Climate affects the performance of cell and metal joints, and cells are corroded by water or moisture. In hot and humid weather, when photovoltaic cells come into contact with water, moisture penetrates into the cell and reduces its efficiency, which results in a decrease in electrical productivity. Studies and research works have investigated the effect of relative air humidity on solar cells, and humidity can be measured using various humidity sensors. A highly sensitive humidity sensor is explained in this chapter. Relative humidity is affected by other climatic variables like temperature; therefore, the effect of these variables should also be included and considered in studies related to PV panels [1].

### 7.1.2 Effect of Humidity on Wind Turbines

It is a known fact that the density of dry air is higher than that of humid air. Humid air means lower density, resulting in reduced power from a wind turbine [2].Wind energy generation plants require humidity control to operate the turbines and other systems efficiently. Insufficient humidity results in damaged components, impaired coating, waste and reduced productivity. Improved humidification control in wind turbines and/or blades improves production output and elevates product quality in the component manufacturing facility. Humidity is the major cause of corrosion, increases the possibility of condensation, can cause service failures and breakdowns, and reduces the life of components.

### 7.1.3 Humidity as Renewable Energy

Scientists in the United States claim to have developed a device that can generate electricity from moisture in the air. The device, based around a thin film of electrically conductive protein nanowires, can produce continuous electrical power for around 20 h before self-recharging. The researchers say that such technology could provide clean energy without the restrictions on location and environmental conditions of other renewable energy solutions such as solar cells [3]. Scientists at Israel's Tel Aviv University reported a new source of renewable energy: the humidity in air, which can generate a voltage using only water and metal, like batteries that can be charged [4]. Inventions and innovations are going on to harness energy from humid weather. US scientists have conducted some experiments to run a toy by harnessing energy from humidity [5].

These facts revealed that atmospheric water vapor could serve as a renewable energy source. In air, atmospheric humidity shares majority of water in vapor form with relative humidity (RH) which is around 5.3 g water/m$^3$ at 22.5°C. This can be utilized as a water source for clean energy generation. Harvesting energy from the atmosphere and environment offers clean power for self-sustained systems. Existing technologies like solar cells, mechanical generators, and thermoelectric devices have specific environmental needs that restrict and limit the continuous production of energy. Since atmospheric moisture is ubiquitous, it can serve as an alternative for clean energy generation.

## 7.2 HUMIDITY SENSORS

There are two basic types of sensors used to measure humidity, the resistive type and the capacitive type. Some researchers have shown that capacitive sensors are more robust to temperature error compared to resistive sensors [6–7]. The capacitive method of humidity detection is one of the most widely used techniques reported in the literature.

The dielectric constant of air is affected by humidity. As humidity increases, the dielectric constant also increases. The permittivity of some atmospheric air, some gases, and many solid materials are functions of moisture content and temperature. Capacitive humidity devices are based on changes in the permittivity of the dielectric material when area and distance between the plates of capacitor are fixed. An increase in the percentage of humidity increases ionic concentration, thereby changing the permittivity, which results in a change in capacitance.

Since the range of humidity is very wide, work is proceeding to measure humidity at different levels. However, the enhancement of sensitivity is still a challenge to humidity measurements over a broad dynamic range. Attempts are being made to use advanced fabrication techniques to develop nanoporous structure sensors for better sensitivity. However, sensitivity improvement is not ensured. Ceramic materials are extensively used to measure humidity. The alumina ceramic material is best suited for fabricating capacitive sensors, since it is highly hydrophilic to water molecules. Other materials such as organic polymers can also be used but have not been shown to be reliable due to low temperature stability and degradation of the film at high humidity. Aluminum oxide ($Al_2O_3$) or alumina possesses the properties of temperature stability, resistance to chemicals, robustness, and rapid response [8].

## 7.2.1 Anodized Capacitive Sensor

This chapter describes development of a capacitive humidity sensor with an anodic film where a change in humidity results in a change in the dielectric constant of the capacitive sensor. The capacitive sensor may be fabricated using the sol-gel technique and anodization method. Since the sol-gel method is time consuming and laborious, and sometimes the sensing films are cracked, the anodization method is preferred to form the porous sensing layer. The sol-gel method is a chemical route where a thin film of alumina is formed on a metal plate using a sol solution prepared using an aluminum sec butoxide precursor. The film is heated at a high temperature of approximately 500°C for formation of the hydrophilic aluminum oxide sensing film. Aluminum oxide, which is mostly reported for humidity sensing, is amorphous in nature, so there is a possibility of degradation of the film, which leads to a drift of the output of the sensor with time. In addition, the film may crack due to the evaporation of vapor molecules during heating. The anodization method is inexpensive and easy, and pore morphology can be controlled by suitably selecting the anodization parameters, namely voltage, time, concentration of electrolytes, and conductivity of the aluminum sheet. Anodic coatings are grouped according to the action of the electrolyte used to obtain the anodic film. Prominent electrolytes are oxalic, phosphoric, and sulfuric acid solutions. When aluminum is anodized in an acid solution, deep pores are formed within the nanometer range, and the length of the porous film may be up to several microns [9–11]. A metal oxide film is formed on both sides of the pure alumina sheet, so there is a bulk metal sandwiched between two oxide surfaces. A capacitive sensor can be formed on each side. The capacitive humidity sensors on each side are connected in parallel so that its value is doubled and the vapor molecules have twice the surface area of the individual sensor for condensation. Hence, the sensitivity can be enhanced. Nanoporous alumina has been used for different applications, including humidity sensors, but the method described in this chapter has increased sensitivity. This chapter focuses on the fabrication of a capacitive humidity sensor with anodized porous alumina. The response characteristics of the sensor were investigated and studied. The sensitivity of the sensor can be enhanced by optimizing the pore morphology of the thin film of $Al_2O_3$ (alumina). To further enhance the sensitivity, two capacitors with identical dimensions on each side of the substrate were fabricated and connected in parallel. The method offers sensitivity enhancement with unaltered sensor dimensions, no waste of the anodized layer on the opposite face of the substrate, and a reduced possibility of electrode shorting, as may occur with capacitive sensors prepared by some of the methods described earlier.

### 7.2.2 Aluminum Oxide as Sensing Film

Aluminum oxide is generally referred to as alumina and is a very cost-effective and widely used material in the field of engineering ceramics. Its key properties include excellent dielectric properties,; it is hard and robust, has good thermal conductivity, is thermally and chemically stable, and has high strength and stiffness [12–15]. The ability of oxidized aluminum to sense humidity is based upon ionic conduction; the presence of water film at the surface lowers sensor impedance due to the increase in ionic conductivity. The capacitance also increases because of the large surface area available for adsorption. Oxidized film is very stable and shows good sensitivity.

### 7.2.3 Anodization

Alumina strips can be anodized in different ways to obtain samples of different pore morphology, that is, voltage, current, area of anodized layer, current density, thickness of strip, distance between electrodes, and electrolyte solutions. Here anodization is done using different solutions as electrolytes. One is oxalic acid solution, and the other is sulfuric acid solution. A platinum electrode served as a cathode. The pretreated alumina strip samples and platinum cathode were separated by a distance of around 3 cm. Two samples were prepared, S1 with 4 wt. % of oxalic acid and $S_2$ with 4 wt% sulfuric acid solution. Both samples were anodized at 30 V for 2 hrs. As the strip was immersed in solution, it anodized on both sides, which helped form two capacitive sensors using alumina as common substrate and electrode. After anodization, the samples were rinsed thoroughly in DI water and then dried in an oven.

## 7.2.4 Pore Morphology

The surface morphology of the anodized alumina samples was studied. Figure 7.1 shows scanning electron microscope (SEM) images of samples S1 and S2. Sample S1 has a predominantly ordered structure of pores, whereas sample S2 did not have a regular periodic porous structure [11].

## 7.2.5 Fabrication of Sensors

Capacitive sensors C1 and C2 fabricated from sample S1 were fabricated on anodized alumina substrates. The silver electrodes on each anodized layer on the alumina substrate were deposited by the screen printing method using a hand screen printer. The printed silver paste was then sintered at 150°C for 3 hrs in an oven. Two electrical connections were taken from

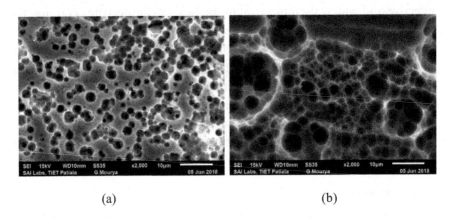

(a)                                          (b)

FIGURE 7.1   SEM images of anodized alumina using (a) oxalic acid solution as electrolyte, and (b) sulfuric acid as electrolyte

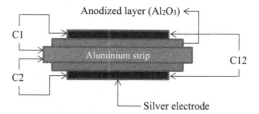

FIGURE 7.2   Schematic diagram of the capacitive sensor

silver electrodes on each side using silver glue. The third connection was taken from the portion of unanodized aluminum substrate common to both parallel plate capacitors C1 and C2.

Figure 7.2 is the schematic diagram of the sensor depicting the structure of the sensor with various layers. C1 and C2 are the two capacitive sensors formed from single structure. The first or innermost layer is the aluminum strip which serves as common electrode. The aluminum oxide layer is the second layer which is porous. The third layer is the silver paste coating for electrode connections.

## 7.2.6  Determination of Response Characteristics of the Sensors

The electrical characteristics were determined by holding the sensors inside the desiccator which contain saturated salt solution. Saturated salt solutions of varying relative humidity were used. The lowest RH (10%) was obtained by placing hot silica gel inside the desiccator, and the highest (99%) humidity was obtained using DI water. Other saturated salt solutions were used to get humidity levels between 10% and 99%. The sensors were shielded properly to guard against stray capacitances. An LCR meter was used to measure capacitance values at different humidity values as well as at different frequencies, and the curve can be plotted as shown in Figure 7.3. LCR meter is an instrument used to measure inductance (L), capacitance (C), and Resistance (R).

It can be observed from Figure 7.3 that the capacitance value increases with the increase in humidity [11]. It has been observed that if the signal frequency is increased, capacitance change with humidity becomes smaller. It is also observed from the curve of Figure 7.4 that the capacitance increases slowly with relative humidity, rising up to 45% RH. The capacitance increases abruptly for higher values of humidity up to 99% RH.

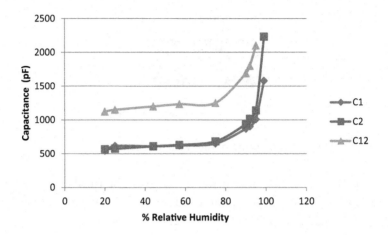

FIGURE 7.3 Variation of capacitance with percentage of relative humidity at 100 kHz

When a sensor is exposed to a certain humidity level, the water molecules in cluster form condense on the porous surface and then in the voids through capillary condensation in the nanopores. As a result, the pores are filled up, causing an increase in the effective dielectric of the sensor. The size of the water clusters depends on the moisture concentration. In saturation conditions, when the pores are completely filled up, the capacitance value reaches the maximum [16].

### 7.2.7 Transient Response

The transient response curve of the sensor C34 with sample S2 obtained with a step change in humidity from 1% to 45% RH is shown in Figure 7.4 [11]. The response and recovery time of the sensor were 151 and 44 s, respectively.

*Repeatability:* The response time is the time taken by the capacitive sensor to change the output from 10% of its maximum value, while the recovery time is the time taken by the sensor to return from 90% output to 10% of its initial value. For real-time applications, these parameters should be as small as possible [17–19]. The repeatability for sensor output for the same humidity change for several cycles is shown in Figure 7.5 [11]. The anodized humidity sensor fabricated here shows improved sensitivity as compared to the humidity sensor with solgel coating reported earlier.

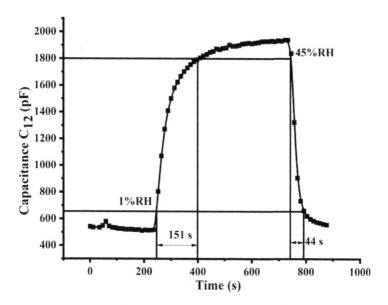

FIGURE 7.4  Transient response of sensor C12

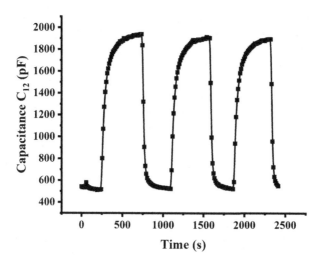

FIGURE 7.5  Dynamic response of sensor C12 to determine the repeatability of the output

## 7.3 CONCLUSION

Humidity sensing is an important parameter in clean energy production, as most of the equipment is exposed to atmospheric air and moisture such as in solar, wind energy, and so on. Humidity affects the performance of

PV cells and wind turbines, which reduces the efficiency of the cell, resulting in a decrease in electrical productivity. Owing to the importance of humidity measurement in these areas, this chapter dealt with the fabrication of a capacitive anodic alumina-based sensor. The characteristics of sensors were examined for 10% to 99% RH using saturated salt solutions of different RH values. It was found that sensors with larger effective area have higher sensitivity. Pore morphology, sensing area, and concentration of electrolytes play important roles in the sensitive detection of humidity. Another feature of the sensors fabricated here consisted of a capacitor on each side of the substrate, hence utilizing both anodized sensing layers on either side of aluminum substrate. This double-sided capacitor structure is easy to fabricate and use, and it enhances sensitivity greatly without affecting the overall size and dimensions. The sensor is capable of withstanding harsh weather conditions due to its benefits of being thermally and chemically stable. Also, the sensor is robust and has high mechanical strength.

## REFERENCES

[1] L. M. Nhut and D. T. T. Linh, "Effects of the relative humidity on the performance of thermoelectric freshwater generator using solar power source," in *5th International Conference on Green Technology and Sustainable Development (GTSD)*, pp. 232–235, Vietnam, 2020.

[2] P. S. Junaidh, A. Vijay and M. Mathew, "Power enhancement of solar photovoltaic module using micro-climatic strategies in warm-humid tropical climate," in *2017 Innovations in Power and Advanced Computing Technologies (i-PACT)*, pp. 1–6, Vellore, 2017.

[3] Salih and Majli, "Relative humidity effect on the extracted wind power for electricity production in Nassiriyah City," *Journal of Fundamentals of Renewable Energy*, vol. 6, no. 1, 2015.

[4] "Nanowire device generates electricity from ambient humidity," *Physics World, Materials for Energy*. https://physicsworld.com/a/nanowire-device-generates-electricity-from-ambient-humidity/. Accessed 3 March 2020.

[5] N. Lavars, "Scientists create an electrical charge using only humidity in the air," New Atlas. https://newatlas.com/energy/electrical-charge-humidity-renewable-energy. Accessed 9 June 2020.

[6] Climate home news, Internet Publishing, 2015. http://www.climatechange news.com.

[7] X. Liu et al., "Power generation from ambient humidity using protein nanowires," *Nature*, vol. 578, pp. 550–554, 2020.

[8] C. K. Chung and C. A. Ku, "Effect of humidity on nanoporous anodic alumina oxide (AAO)," *World Journal of Nanoscience and Nanotechnology*, vol. 1, p. 1003, 2019.

[9] Y. Wang, S. Hou, T. Li, S. Jin, Y. Shao, H. Yang, D. Wu, S. Dai, Y. Lu, S. Chen and J. Huang, "Flexible capacitive humidity sensors based on ionic conductive wood-derived cellulose nanopapers," *ACS Applied Materials & Interfaces*, vol. 12, 37, pp. 41896–41904, 2020.

[10] S. Kano and H. Mekaru, "Nonporous inorganic nanoparticle-based humidity sensor: Evaluation of humidity hysteresis and response time," *Sensors*, vol. 20, no. 14, p. 3858, 2020. https://doi.org/10.3390/s20143858

[11] A. Salamat and T. Islam, "Fabrication of anodized porous alumina relative humidity sensor with improved sensitivity," *Instrumentation Science and Technology*, vol. 48, no. 2, 2019. doi:10.1080/10739149.2019.1662803

[12] T. Islam, A. T. Nimal, U. Mittal and M. U. Sharma, "A micro interdigitated thin film metal oxide capacitive sensor for measuring moisture in the range of 175–625 ppm," *Sensors and Actuators B: Chemical*, vol. 221, pp. 357–364, 2015.

[13] L. Kumar, T. Islam and S. C. Mukhopadhyay, "Sensitivity enhancement of a PPM level capacitive moisture sensor," *Electronics*, vol. 6, no. 41, 2017.

[14] Z. H. Zargar and T. Islam, "Thin film porous alumina-based cross-capacitive humidity sensor," *IEEE Transactions on Instrumentation and Measurement*, vol. 69, no. 5, pp. 2269–2276, 2019.

[15] T. Islam, "Advanced interfacing techniques for the capacitive sensors," in *Advanced Interfacing Techniques for Sensors-Book*, pp. 73–109, Springer, Cham, 2017.

[16] T. Islam and M. Z. U. Rahman, "Investigation of the electrical characteristics on measurement frequency of a thin-film ceramic humidity sensor," *IEEE Transactions on Instrumentation and Measurement*, vol. 65, no. 3, pp. 694–702, 2015.

[17] M. Y. Cho, S. Kim, I. S. Kim, E. S. Kim, Z. J. Wang, N. Y. Kim, S. W. Kim and J. Min Oh, "Humidity sensor systems: Perovskite-induced ultrasensitive and highly stable humidity sensor systems prepared by aerosol deposition at room temperature," (Adv. Funct. Mater. 3/2020), *Advanced Functional Materials*, p. 1907449, 2019.

[18] O. Arghese, D. Gong, M. Paulose, K. Ong, C. Grimes and E. Dickey, "Highly ordered nanoporous alumina films: Effect of pore size and uniformity on sensing performance," *Journal of Materials Research*, vol. 17, no. 5, pp. 1162–1171, 2007. doi:10.1557/JMR.2002.0172

[19] T. Islam, Z. H. Zargar and M. Rehman, "A novel humidity sensor based on the extension of Thompson and Lampard theorem," *IEEE Transactions on Electron Devices*, vol. 62, no. 12, pp. 4237–4241, 2015.

# CHAPTER 8

# Energy-Efficient Optimized Routing Techniques in an IoT-Enabled Intelligent Traffic Management System

Piyush Agarwal, Sachin Sharma, Priya Matta

CONTENTS

## 8.1 INTRODUCTION

Industrialization has resulted in growth in the population of metropolitan centers in many countries. Increases in population have a direct

DOI: 10.1201/9781003218760-8

or indirect impact on road traffic. The existing traffic management infrastructure is unable to properly control road traffic due to a lack of resources. This inefficiency causes concerns such as traffic congestion, which increases travel time, as well as a variety of health and environmental hazards. According to a report by the US Federal Highway Administration, there are three main causes of traffic congestion: i) traffic-changing events (such as traffic resource maintenance, weather conditions, and traffic incidents), ii) user demand, and iii) resource physical features (such as capacity and traffic control devices). Urbanization raises transportation demand, which necessitates upgrading resources to meet the increased demand. Events such as resource upgrades, weather conditions, and other occurrences have a direct impact on traffic flow and cause congestion.

The UK's Center for Economics and Business Research (CEBR) looked at the direct and indirect costs of congestion in the American, British, German, and French scenarios. In the study, the direct cost is defined as the time and fuel wasted due to traffic congestion, while the indirect cost is defined as business loss. The cost of congestion was predicted up to 2030 in this study. From 2013 to 2030, the overall cost of congestion in these countries is estimated to be around $4.4 trillion [1]. An alternative technical approach for reducing traffic congestion could be an IoT-enabled intelligent traffic management system (ITMS), which will improve traffic flow, quality of life, safety, and security. An IoT-enabled ITMS integrates data gathered from many sources such as roads, cars, or people with a computer engine via a communication channel to solve the challenges of traditional traffic management systems. The IoT-enabled ITMS employs machine learning and deep learning techniques and algorithms to provide users with energy-efficient apps that reduce congestion, improve system efficiency, improve safety, and lower travel costs. The node in the Internet of Things transmits information regarding traffic conditions, route discovery, and other traffic information. This transmission mechanism uses all of the node's energy. Almost all nodes operate on a restricted power source because they are powered by batteries, so optimizing battery usage in IoT is critical. Route discovery packets are transmitted from one to another in IoT-enabled ITMS and consume the most energy. The route of a packet can be determined using machine learning approaches for energy efficiency.

## 8.2 LITERATURE REVIEW

Chavhan et al. [2] proposed a TMS using static and mobile agents. Based on the mobility pattern, the agents are classified into static and mobile agents for traffic information gathering, analysis, and resource allocation at different locations of the city and can take the decision. Mobile agents collect and share information based on traffic speed and density, historical data, and information related to resources using the emergent intelligence technique. The static agent uses this information for monitoring and predicting the traffic patterns in each zone and region. This predicted information is used to identify the optimal routes to divert traffic and to reduce congestion for smooth traffic flow.

A route suggestion protocol is proposed by Ahmed et al. [3] that will work for both equipped and non-equipped vehicles and increases the congestion index accuracy. The proposed model provides route suggestions considering the travel time and congestion on the road segment.

A centralized traffic manager is proposed in [4] to address the problem of congestion. An equation is proposed to predict traffic congestion using the number of vehicles on each street and the travel time. The information on traffic congestion is used to suggest the number of alternative routes to balance the traffic.

An algorithm is proposed by Tian et al. [5]. The proposed algorithm is initialized based on the current traffic situation on the road segment. To map the route selection, a strategy is applied for auto-correction in backtracking and uses the iterative results. The proposed algorithm can calculate the cost of the selected path. This cost includes fuel consumption, road conditions, and other factors. To obtain the optimal solution, it uses the iterative process to adjust the current iteration step according to the current result.

An infrastructureless distributed vehicle traffic management system is proposed by Akabane et al. [6] to detect and avoid congestion based on a vehicular ad-hoc network (VANET). The proposed system consists of three layers: i) the environment sensing layer, ii) the knowledge generation and distribution layer, and iii) the knowledge consumption layer. The first layer of the proposed method chooses the vehicle on the road based on the rank of the vehicle calculated by the egocentric betweenness metric and the radio propagation model. The congestion level on the road segment is identified by the knowledge generation and distribution layer. The vehicle chosen by the previous layer transmits information about the congestion and speed of the road to other vehicles. Depending on the popularity of the road segments and using the proposed shortest path entropy-based algorithm, the system suggests an alternative route to the destination.

A service named Re-RouTE was proposed by Guidoni et al. [7]. The proposed system works in four steps: i) location information, ii) network representation, iii) network classification, and iv) route suggestion. In the first step, the current location and destination are shared by every vehicle before the start of the journey. In the second step, the system generates the weighted graph of the road network, where weight represents the density of the different road segments of the city. The weighted graph generated in the previous step is analyzed by the network classification layer to classify the road segment as congested or non-congested. The optimized path is suggested by the route suggestion module.

Tomar et al. [8] proposed logistic regression with fuzzy-logic-based work for intelligently suggesting the alternative path. This alternative path was selected based on the real-time traffic information, distance to be covered, and factors like weather conditions and road location. In the proposed machine learning-based system, logistic regression is used to get the details of each possible path. This set of all possible paths is provided as input to fuzzy logic to get the best possible path.

Traffic congestion aware route assignment (TCARA) is proposed in [9] using a predictive congestion model to reduce traffic congestion. The proposed algorithm is based on the A* algorithm with the same complexity as Dijkstra's algorithm. The pressure on the road link is used to identify the congestion on the road and hence the shortest route to the destination.

Motallebi et al. [10] proposed a routing algorithm for TMS. The algorithm receives the route request for the given source and destination. The algorithm can take three types of input: i) source and destination of the vehicle that is shared by it at the start of the trip, ii) time taken by the vehicle to travel the road link, and iii) updated routes. When a vehicle starts its trip, it shares its route source and destination; based on the input received, the algorithm generates the route. The average travel time of the vehicles for road links is taken as input to get the traffic conditions on the road links. Once the vehicle crosses the road link, that road link is removed from the route to generate the route in the future. The algorithm maintains the reservation count for all the road links, which is used to predict the traffic load on the road links. A local detour algorithm is used to detect intersections in the suggested route.

To transmit the data from a node to a server, Wang et al. [11] proposed a method that divides sensors into different sectors or clusters of the same size, each cluster with a cluster head that is responsible for inter cluster communication. Each cluster member identifies the optimum routing algorithm to send data to the cluster head, and, using the greedy algorithm, the cluster head forms a chain for intercluster communication.

A routing algorithm is proposed by Khirji et al. [12] that combines the localization and clustering method to reduce the energy consumption of the network. To determine the position of the nodes, a localization-based received signal strength indicator (RSSI) algorithm is used. To balance the energy consumption among the sensors of the nodes, a fuzzy-based unequal clustering algorithm is used.

Li et al. [13] proposed an energy-efficient load-balancing ant colony-based route optimization algorithm for wireless sensor network (WSN). The goal of the proposed algorithm is to balance energy consumption among the nodes, to increase the lifetime of the network, and to reduce the route discovery time with the constraint of the energy supply. To speed up the task of identifying the effective route, a pseudo-random algorithm is used considering the energy consumption. The energy level of the node, information related to the path length, and results of the prolonged network lifetime are considered in the trail update procedure. Based on the greedy expected cost of energy, a heuristic approach is used, which further reduces the route establishment time.

A grid-based routing algorithm was proposed in [14]. The grid controller was chosen according to the residual energy of the nodes within the grid. The information from the source to the sink is transmitted by the grid coordinator to reduce the energy consumption in the routing process. Fuzzy rules are developed for the efficient and effective deployment of the node, formation of the cluster, selection of the cluster head, routing, and energy analysis.

Focusing on energy consumption, Safara et al. [15] proposed a priority-based energy-efficient routing algorithm based on low power and a lossy network routing algorithm to increase robustness. The algorithm uses the time division multiple access (TDMA) model to send the data considering network traffic, audio, and image data.

An ant colony optimization routing technique for energy-efficient routing for MANET is proposed by Malar et al. [16]. In the proposed algorithm, the hope is selected considering constraints such as residual energy, number of packets in the channel, and movement.

Elsemany et al. [17] proposed an energy-efficient clustering and hierarchical routing algorithm. The algorithm aims to increase the lifetime of the network. Using three-layered architecture, the load of the cluster head is minimized, and the procedure of cluster head selection is randomized. Intra-cluster communication is established using multi-hop transmission.

To enhance network performance, Thangaramya et al. [18] worked on clustering and proposed a coevolutionary neural network and fuzzy rule approach for routing based on clusters. The proposed approach considered factors such as residual energy of CH, distance between the sink node and the cluster head as the distance between the node and cluster head, and degree of the cluster head for effective utilization of the energy to increase the lifespan of the network.

Tang et al. [19] use the Dempster–Shafer evidence theory for an energy-efficient and reliable routing algorithm based on the residual energy of the neighboring node, traffic, closeness to the shortest path, and so on. An attribute with three indexes is established, and, using the entropy weight method, the weight of these indices is determined. For the selection of the next step, the basic probability assignment function is established, and each index value is fused using the Dempster–Shafer theory.

To reduce power consumption for transmitting packets in the network in VANET, [20] proposes a greedy ad-hoc on-demand distance vector-based routing protocol. In the communication flows, the total power consumed by the participating node is considered the selecting parameter of route formation.

## 8.3 OVERVIEW OF IoT-ENABLED ITMS

The IoT offers technologies that are both cost effective and simple to implement for route optimization. These devices are simple to install and do not require any extra modifications to the car. IoT is mostly utilized for real-time vehicle tracking and tracing, particularly by logistics-based businesses [21]. Large amounts of data may be acquired with IoT devices to increase logistical efficiency even more. Better vehicle routing, for example, can make the operation more efficient. Individual vehicles or roadside IoT sensors can be deployed in order for the system to obtain real-time vehicle information. This real-time data can be analyzed on the server to determine the level of congestion in various parts of the city. The optimal path from source to destination might be given based on the level of congestion in various locations. The vehicle's real-time information is obtained using IoT sensors. A safety camera, radio frequency

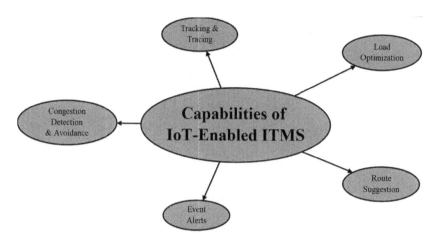

FIGURE 8.1    Capabilities of IoT-enabled ITMS

identification reader, ultrasonic sensor, motion sensor, proximity sensor, passive infrared sensor, sound detection sensor, radar sensor, and pressure sensor are some of the sensors that can be utilized for this purpose [22]. As indicated in Figure 8.1, the obtained data can be used to give the following capabilities.

- *Tracking and Tracing*: Tracking and tracing provide the exact location of the vehicle using a GPS device, surveillance camera, or IoT sensors. The location of the vehicle helps in analyzing the driving pattern of the driver, and the number of vehicles on the particular road segment can be calculated. Law enforcement authorities can track vehicles involved in illegal activities.

- *Load Optimization:* Every road segment has a threshold capacity to allow vehicles for free flow. As the number of vehicles crosses the threshold capacity, the chances of congestion in that particular road segment increase. To prevent vehicles from entering the road segment, they can be rerouted to their destination through another optimized path.

- *Route Suggestion:* The shortest route to the destination reduces travel time and fuel consumption. The IoT-based ITMS suggests a route from the current location of the vehicle to the destination considering traffic congestion on different road segments, events, weather conditions, road conditions, and other factors affecting the traffic flow and the environment.

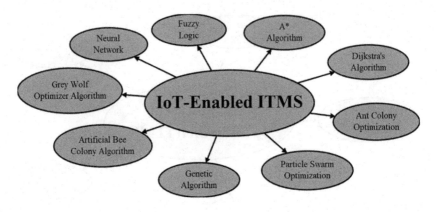

FIGURE 8.2  Energy-efficient optimized routing techniques in IoT-enabled ITMS

- *Event Alerts:* To reduce congestion, the IoT-based ITMS can provide information about events such as accidents and terrorist attacks to the traveler and hence can suggest an alternative route.

- *Congestion Detection and Avoidance:* Using sensors or a surveillance camera, and with pre-decided conditions of congestion, congestion can be predicted before it occurs, and the necessary steps such as altering the path of the vehicle can be taken.

## 8.4 IMPORTANCE OF ENERGY-EFFICIENT OPTIMIZED ROUTING TECHNIQUES

Route optimization handles tough problems such as vehicle routing to alleviate traffic congestion on the road, commuter travel, public transportation trips, and delivery-based businesses. Route optimization does not always provide the shortest route from point A to point B; the route provided by this technique is dependent on a number of factors, including traffic congestion on various road segments, weather conditions, and the distance and travel time of various routes. Because identifying the optimal route necessitates the use of energy to transport data from one node to the next, it is critical to maximize energy efficiency during processing for optimal route identification. Energy utilization can be affected by using an efficient routing path for the packet to reach its destination node and increase the lifespan of the network. Different techniques that can be used in an IoT-based ITMS are shown in Figure 8.2.

- *Neural Network:* A neural network consists of neurons with connected weights. Due to its parallel processing of large-scale data with the storage of distributed information, self-learning can be used for optimized route-finding algorithms. The optimized objective function is the sum of the distance function and the collision energy. The optimal path can be set by the iterative path point using the equation of motion [23]. The neural network can be used for the process of identifying the optimized route to reduce the consumption of energy. [18] trained a neural network using past and current communication data.

- *Fuzzy Logic:* Fuzzy logic is a soft computing method that is used to handle uncertainty. The degree of membership or value of uncertainty falls between 0 and 1, where 1 means that the member is the part of the set and 0 denotes a non-member of the set. Road traffic is uncertain and unpredictable, so it can be incorporated to decide on the preferred path [8]. To increase the energy efficiency of the network, fuzzy logic is used by [18] to organize the cluster and network with features such as radius of the cluster and non-probabilistic cluster head selection.

- *A\* Algorithm:* The A\* algorithm uses a heuristic function for the calculation and comparison of the neighboring node values. For an efficient solution, the algorithm selects the node with the smallest value; this process continues until the target node is selected. Load on the road segments is used to calculate the weight in the A\* algorithm [9–10].

- *Dijkstra's Algorithm:* To get the shortest path from source to destination, Dijkstra's algorithm compares the values of the nodes and selects the node with a lower value. The major disadvantage of the algorithm is that to get the shortest path, it traverses all the nodes.

- *Ant Colony Optimization (ACO):* ACO is a probabilistic technique for solving the optimal path problem for a given graph. The working phenomenon of ACO is similar to that of ants searching for food. The artificial ants work as the simulating agents move around the space with some parameter value based on the model to locate the optimal solution. By recording the positions with solution quality, at a later stage, other ants can locate a better solution. In [13], ACO uses the probabilistic transition rule to select the next node for the energy-efficient and load-optimized routing in WSN. As more ants travel by the shortest path, the attention level rises, and the shortest way is discovered through repetition iteration. ACO is capable of global optimization and is simple to implement.

- *Particle Swarm Optimization (PSO):* PSO is a method that iteratively optimizes the solution in a feasible region of a given solution. This algorithm is based on the movement of birds in the group and sharing the information among the individuals in it. Similarly, in the algorithm, iteratively, to get the optimum solution, individuals update themselves with the help of their historical solutions as well the global optimal solutions.

- *Genetic Algorithm:* For natural selection of Darwin's biological evolution theory, a simulation solution is provided by a genetic algorithm. Initially, the potential solution for the population is coded, followed by the calculation of the adaptive individual value. Individuals are eliminated by performing selection, cross-over, and mutation using the principle of survivability of the fittest. Repeating in the iteration, the algorithm reaches the best individual.

- *Artificial Bee Colony (ABC) Algorithm:* An ABC algorithm is an optimized and efficient metaheuristic intelligent swarm method based on the behavior of honeybees. To optimize the clustering method, an energy-efficient routing algorithm for data transmission between the cluster head and the base station [24] uses an ABC algorithm.

- *Grey Wolf Optimizer (GWO) Algorithm:* A GWO algorithm simulates the hunting and leadership hierarchy of grey wolves. Alpha, beta, and delta wolves have optimum knowledge of the prey; similarly, GWO stores the three best optimum solutions and discards the others; this information is used to update their positions. Manshahia et al. [25] implemented GWO for optimizing the fitness function that is based on the residual energy and throughput of the IoT network.

## 8.5 FEATURES OF IoT-ENABLED ITMS

The traffic management system will be more efficient and reduce energy consumption using IoT technologies and route optimization techniques. Figure 8.3. depicts some of the features of the technologies discussed in the previous section.

### 8.5.1 Efficiency

a. *Fuel Efficiency:* Route optimization identifies the best route to the destination that has less traffic and is distance effective as compared to other routes. The route has less traffic, which decreases fuel consumption. The distance travelled may be more as compared to other routes since the system may have considered factors such as weather conditions, road conditions, traffic conditions, travel time, fuel consumption, and so on.

b. *Travel Time:* The route that is selected by the system is optimized; also, the route selection considers the factors like time taken, distance traveled, and so on. Therefore, the selected route is time efficient, or, in other words, one can say that travel time is lowest for the selected route.

## 8.5.2 Competency

c. *Scalability:* Route planning for businesses, like logistics, involves consideration of the area, availability, traffic conditions, weather, and much more. It will take a long time to manually plan the routes for such a business. With a route optimization system, this time and manpower used for route planning can be reduced. Route optimization systems work efficiently with the increase in deliveries per day and with the increase in the area of delivery.

d. *Accuracy:* A route optimization system considers all complexities while finalizing the route, which reduces the chances of error and accidents. On the other hand, when the route is planned manually, there are lots of chances for error, and as the complexity increases, the chances of error also increase.

e. *Flexibility:* There are chances of non-availability of any vehicle; the route optimization system can provide an alternative plan with the available vehicles considering the delivery time.

Features of IoT-enabled ITMS can be categorized into five categories: efficiency, competency, safety, resource optimization, and business insights (Figure 8.3). Fuel efficiency and reduced travel time come under the category of efficiency, whereas scalability, accuracy, flexibility, and robustness are under competency. Under resource optimization, there are three features: efficient resource allocation, fleet management, and dynamic re-routing. Business insights has four feature: better customer experience, delivery prioritization, real-time updates, and driver morale

f. *Robustness:* While using a route optimization system for generating a route plan, the operator doesn't need to know the geography of the

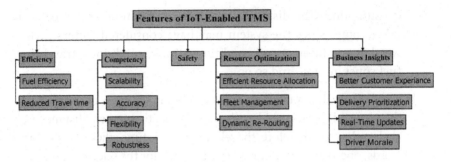

FIGURE 8.3    Features of IoT-enabled ITMS

city. Also, as an individual, a route optimization system can suggest the best route from the source to the destination even if the driver is not familiar with the city.

**Safety**: As the traveler spends more time on the road, the risk of threats also increases. A route optimization system not only reduces the travel time but can also suggest a route that includes the least number of intersections and turns to decrease the chances of accidents and increase the safety of the traveler.

### 8.5.3 Resource Optimization

g. *Efficient Resource Allocation:* In the logistics and e-commerce business, route optimization systems help in optimizing the number of trips, which reduces the requirements for vehicles.

h. *Fleet Management:* A route optimization system can handle fleet management operations like trip schedules with a suitable number of stops in one trip.

i. *Dynamic Re-Routing:* Electric/autonomous vehicles can communicate with the outer world. Route optimization system can make use of this feature to get real-time traffic scenarios to generate alternative routes to the destination. For logistics, it can provide an alternative route to the delivery location with on-time delivery.

### 8.5.4 Business Insights

j. *Better Customer Experience:* The goal of logistics and cab service-like business is to make the customer experience better. In the logistics business, for positive customer feedback, on-time delivery is necessary. Many factors affect the delivery time. One of the main factors that affect the delivery is traffic congestion and the route that is followed by the delivery person to deliver the product. Route optimization techniques not only suggest the route but can also suggest the mode of transportation at different stages of the delivery, keeping to the deadline of the delivery, cost, and many other factors.

k. *Delivery Prioritization:* The route can be generated such that if there are items whose delivery time is before the others and distance is great, they can be delivered early. This route optimization system makes sure that other deliveries will not be affected.

l. *Real-Time Updates:* The route can update based on the real-time situation of the traffic at different places. The real-time situation of the traffic can be gathered using sensors placed at the roadside or through vehicles.

m. *Driver Morale:* The route suggested by the route optimization algorithm helps drivers reach their destination on time. This boosts morale as well as increasing the driver's confidence, and they can concentrate on the path.

## 8.6 CHALLENGES OF IoT-ENABLED ITMS

The security and privacy of users is a fundamental concern in wireless communication. In addition, in an IoT-based system, wireless communication is used the majority of the time. Because this is a new field, certain questions about feasibility have arisen. The difficulties of security, privacy, and feasibility are highlighted in this section in relation to IoT-based ITMS (Figure 8.4).

• *Confidentiality:* Confidentiality ensures that data will be accessed by authorized users only. In route optimization, the route details, delivery status, and vehicle location should be confidential and should be accessed only by the required users. The unauthorized user should not be able to look at or modify the data of the users. Therefore, while developing a route optimization system, cryptographic methods should be implemented [26].

• *Integrity:* In a route optimization system, data integrity should be present, as it guarantees the delivery of exact and accurate

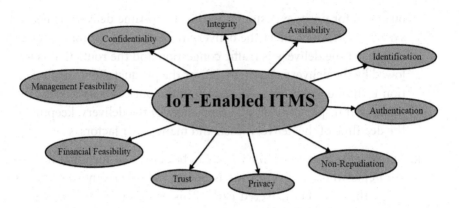

FIGURE 8.4 Challenges in IoT-enabled ITMSs

information or data. It also ensures that there is no unauthorized access while transferring the data. Incorrect information may lead the driver to the wrong route or result in congestion instead of making the road congestion free. Integrity also ensures that there is no intervention during the communication. Therefore, ITMS should include security mechanisms [27].

- *Availability:* The route optimization system should always be available to the citizens or authorized users. The system should safeguard itself from the threats like denial-of-service attacks that may make the system unavailable. Maheswari et al. [28]

- *Identification:* A route optimization system should ensure that data should be accessed or transferred by the authorized device or vehicle. Since in route optimization, there is no limit to the vehicles, the proper method should be available for identification of the vehicle.

- *Authentication:* In route optimization, it is important that the user receive valid messages; otherwise, a message may misguide the user, which will result in traffic congestion, larger travel distance, and many other problems. To ensure the validity of the message, authentication can be implemented.

- *Non-Repudiation:* Non-repudiation refers to a service that provides proof of the origin and integrity of the data. The messages sent by servers, vehicles, or communicable devices cannot refute it at a later point in time. To avoid malicious threats, non-repudiation must be implemented into the system [22].

- *Privacy:* The data relating to the identity of the vehicle and the user is personal and important and should not be received by an unauthorized person. The security of the identity of the vehicle and the user can be provided by implementing privacy service in the system. [29].

- *Trust:* To accomplish trust during the transmission of data or messages, the security and communication features discussed previously should be implemented. A trustworthy approach must be developed to achieve trust in ITMS [16].

- *Financial Feasibility:* The route optimization system involves vehicles with the ability to communicate with other vehicles as well as with infrastructure; due to this, the cost of the vehicle will increase. For real-time updates, the system requires sensors and communication devices to communicate with the vehicles as well as the infrastructure. The use of these types of equipment will increase the one-time as well maintenance costs of the system.

- *Management Feasibility:* To effectively operate the system, a skilled professional is required. To fulfill this requirement, the organization either must hire a professional or train the existing personnel.

## 8.7 RESULTS, ANALYSIS, AND DISCUSSION

In this section, the performance of the various proposed models in the related work is discussed. In the energy-efficient, load balancing, ant-based routing algorithm proposed by Li et al. [13], the energy consumed by the cluster head is 48 J, and the energy consumed by other nodes in the cluster is 23 J when tested for approximately 4 hours. The total energy consumed is about 78 J with three nodes in the cluster. The energy consumes is about 0.05% of the initial energy. The energy consumption of 200 nodes in energy-efficient grid-based routing using fuzzy rules [14] with an initial energy of 0.5 is approximately 0.07 J, with 200 rounds and a packet size of 4,000 bits. [15] uses supplementary nodes due to which the overall energy consumption is reduced; also, as the number of nodes in the network increases, the energy consumption also decreases. The average energy consumption of the nodes for ten rounds is approximately 0.05 J [17].

A comparative analysis of the recent works on the optimized routing algorithm for a traffic management system and energy-efficient routing algorithm are depicted in Table 8.1 and Table 8.2, respectively.

TABLE 8.1 Comparative Analysis of Route Optimization Techniques in a Traffic Management System

| Reference | Technology/ Algorithm Used | Architecture | Features | | | | Limitations/Future Scope |
|---|---|---|---|---|---|---|---|
| | | | Optimized Route | Load Balancing | Track and Trace | Intersection Free route | |
| [2] | V2V and V2I communication | Hierarchical | ✓ | ✗ | ✗ | ✗ | In the proposed system, there are chances of congestion in the neighboring region in the case when there is very high traffic density in the neighboring region. |
| [3] | V2V and V2I communication | Centralized | ✓ | ✗ | ✗ | ✗ | The number of vehicles on the road can be identified using sensors or video surveillance cameras more accurately. |
| [4] | Route management | Centralized | ✓ | ✗ | ✗ | ✗ | With the increase in the number of vehicles, the load on the server will increase. There is a risk of one-point failure. |
| [5] | Genetic algorithm | - | ✓ | ✗ | ✗ | ✗ | In the proposed algorithm, weather is not considered to calculate the cost of the selected path. |
| [6] | VANET | Infrastructure less | ✓ | ✓ | ✗ | ✗ | Newly discovered routes by the vehicle may not be optimized, as the vehicle does not have the information from other cells. |
| [7] | VANET | Centralized | ✓ | ✓ | ✓ | ✗ | The proposed algorithm is for traffic prediction and for route suggestion, the method could act proactively. |

| | | | | | | |
|---|---|---|---|---|---|---|
| [8] | Logistic regression, fuzzy logic | - | ✓ | ✗ | ✗ | ✗ | A smartphone application can be provided that uses the proposed method. It is integrated with Google Maps to guide the driver to reach the destination. |
| [9] | A* algorithm | - | ✓ | ✓ | ✗ | ✗ | The cycle of traffic lights can be incorporated with the model to enhance the traffic congestion model. |
| [10] | A* algorithm | - | ✓ | ✓ | ✗ | ✓ | The travel time for different categories of vehicles on the same road linked with the same traffic condition will differ. The algorithm can be updated, which will consider this issue. |

TABLE 8.2  Comparative Analysis of Energy-Efficient Route Optimization Techniques

| S. No | Authors | Architecture/ Technique | Algorithm | Features | | | | |
|---|---|---|---|---|---|---|---|---|
| | | | | Optimized path | Energy efficient | Balanced Energy consumption | Network load optimization | Scalable network |
| 1 | Wang et al. [11] | Cluster | Greedy algorithm | ✓ | ✓ | ✗ | ✗ | ✗ |
| 2 | Khirji et al. [12] | Fuzzy logic, clustering | Localization-based RSSI algorithm | ✓ | ✓ | ✓ | ✗ | ✗ |
| 3 | Li et al. [13] | - | Ant colony optimization | ✓ | ✓ | ✓ | ✗ | ✗ |
| 4 | Logambigai et al. [14] | Grid-based clustering | Fuzzy rules | ✓ | ✓ | ✗ | ✗ | ✗ |
| 5 | Safarra et al. [15] | TDMA | - | ✓ | ✓ | ✗ | ✗ | ✗ |
| 6 | Malar et al. [16] | - | Ant colony optimization | ✓ | ✓ | ✗ | ✓ | ✗ |
| 7 | Elsemany et al. [17] | Cluster, hierarchical | Low-energy adaptive clustering hierarchy-based algorithm | ✓ | ✓ | ✓ | ✓ | ✓ |
| 8 | Thangaramya et al. [18] | Convolutional neuro network, fuzzy logic, clustering | - | ✓ | ✓ | ✗ | ✗ | ✗ |
| 9 | Tang et al. [19] | Dempster–Shafer evidence theory | - | ✓ | ✓ | ✗ | ✗ | ✗ |
| 10 | Baker et al. [20] | Ad-hoc on-demand distance vector | - | ✓ | ✓ | ✗ | ✗ | ✗ |
| 11 | Wang et al. [21] | - | ABC | ✓ | ✓ | ✗ | ✗ | ✗ |

## 8.8 CONCLUSION

The increase in traffic on roads leads to traffic congestion. Due to the inefficiency of the traditional traffic management system to manage traffic, it is required to upgrade the system using new technologies. As discussed in this chapter, an IoT-enabled ITMS can manage the current traffic load with various techniques and algorithms. Energy is a major challenge for IoT-based systems; therefore, it is important to use energy-efficient techniques. This chapter discussed the capabilities of the IoT-enabled ITMS and some of the technologies and algorithms that can be used with IoT-enabled ITMS to increase the efficiency of the system and provide an optimized route to the road user. The techniques discussed in this chapter can provide energy-efficient solutions. After the study on features and challenges of the discussed techniques in IoT-enabled ITMS, the chapter covered a summary of the recent work done in energy-efficient route optimization in ITMS. Following a discussion of the features and problems of the methodologies covered in IoT-enabled ITMS, the chapter concluded with a synopsis of recent work in energy-efficient route planning in ITMS.

## REFERENCES

[1] D. L. Guidoni, Guilherme Maia, Fernanda S. H. Souza, Leandro A. Villas and Antonio A. F. Loureiro, "Vehicular traffic management based on traffic engineering for vehicular ad hoc networks," *IEEE Access*, vol. 8, pp. 45167–45183, 2020.

[2] Suresh Chavhan and Pallapa Venkataram, "Prediction based traffic management in a metropolitan area," *Journal of Traffic and Transportation Engineering (English Edition)*, vol. 7, no. 4, pp. 447–466, 2020.

[3] Muhammad Jamal Ahmed, Saleem Iqbal, Khalid M. Awan, Kashif Sattar, Zuhaib Ashfaq Khan and Hafiz Husnain Raza Sherazi, "A congestion aware route suggestion protocol for traffic management in internet of vehicles," *Arabian Journal for Science and Engineering*, vol. 45, no. 4, pp. 2501–2511, 2020.

[4] Jorge Luis Zambrano-Martinez, Carlos T. Calafate, David Soler, Lenin-Guillermo Lemus-Zúñiga, Juan-Carlos Cano, Pietro Manzoni and Thierry Gayraud, "A centralized route-management solution for autonomous vehicles in urban areas," *Electronics*, vol. 8, no. 7, pp. 722, 2019.

[5] Yuefei Tian, Wenbin Hu, Bo Du, Simon Hu, Cong Nie and Cheng Zhang, "IQGA: A route selection method based on quantum genetic algorithm-toward urban traffic management under big data environment," *World Wide Web*, vol. 22, no. 5, pp. 2129–2151, 2019.

[6] Ademar T. Akabane, Roger Immich, Luiz F. Bittencourt, Edmundo R. M. Madeira and Leandro A. Villas, "Towards a distributed and infrastructureless vehicular traffic management system," *Computer Communications*, vol. 151, pp. 306–319, 2020.

[7] Daniel L. Guidoni, Guilherme Maia, Fernanda SH Souza, Leandro A. Villas and Antonio A. F. Loureiro, "Vehicular traffic management based on traffic engineering for vehicular ad hoc networks," *IEEE Access*, vol. 8, pp. 45167–45183, 2020.

[8] Anurag Singh Tomar, Mridula Singh, Girish Sharma and K. V. Arya, "Traffic management using logistic regression with fuzzy logic," *Procedia Computer Science*, vol. 132, pp. 451–460, 2018.

[9] Sadegh Motallebi, Hairuo Xie, Egemen Tanin and Kotagiri Ramamohanarao, "Traffic congestion aware route assignment," in *11th International Conference on Geographic Information Science (GIScience 2021)-Part I. Schloss Dagstuhl-Leibniz-Zentrum für Informatik*, Poland, 2020.

[10] Sadegh Motallebi, Hairuo Xie, Egemen Tanin, Jianzhong Qi and Kotagiri Ramamohanarao, "Streaming route assignment for connected autonomous vehicles (systems paper)," in *Proceedings of the 27th ACM SIGSPATIAL International Conference on Advances in Geographic Information Systems*, pp. 408–411, Chicago, IL, 2019.

[11] Jin Wang, Yu Gao, Wei Liu, Arun Kumar Sangaiah and Hye-Jin Kim, "Energy efficient routing algorithm with mobile sink support for wireless sensor networks," *Sensors*, vol. 19, no. 7, pp. 1494–1497, 2019.

[12] Sabrine Khriji, Dhouha El Houssaini, Ines Kammoun, Kamel Besbes and Olfa Kanoun, "Energy-efficient routing algorithm based on localization and clustering techniques for agricultural applications," *IEEE Aerospace and Electronic Systems Magazine*, vol. 34, no. 3, pp. 56–66, 2019.

[13] Xinlu Li, Brian Keegan, Fredrick Mtenzi, Thomas Weise and Ming Tan, "Energy-efficient load balancing ant based routing algorithm for wireless sensor networks," *IEEE Access*, vol. 7, pp. 113182–113196, 2019.

[14] Rajasekar Logambigai, Sannasi Ganapathy and Arputharaj Kannan, "Energy-efficient grid-based routing algorithm using intelligent fuzzy rules for wireless sensor networks," *Computers & Electrical Engineering*, vol. 68, pp. 62–75, 2018.

[15] Fatemeh Safara, Alireza Souri, Thar Baker, Ismaeel Al Ridhawi and Moayad Aloqaily, "PriNergy: A priority-based energy-efficient routing method for IoT systems," *The Journal of Supercomputing*, pp. 1–18, 2020.

[16] A. Malar, M. Kowsigan, N. Krishnamoorthy, S. Karthick, E. Prabhu and K. Venkatachalam, "Multi constraints applied energy efficient routing technique based on ant colony optimization used for disaster resilient location detection in mobile ad-hoc network," *Journal of Ambient Intelligence and Humanized Computing*, vol. 12, no. 3, pp. 4007–4017, 2021.

[17] Eyman Fathelrhman Ahmed Elsmany, Mohd Adib Omar, Tat-Chee Wan and Altahir Abdalla Altahir, "EESRA: Energy efficient scalable routing algorithm for wireless sensor networks," *IEEE Access*, vol. 7, pp. 96974–96983, 2019.

[18] K. Thangaramya, Kanagasabai Kulothungan, R. Logambigai, M. Selvi, Sannasi Ganapathy and Arputharaj Kannan, "Energy aware cluster and neuro-fuzzy based routing algorithm for wireless sensor networks in IoT," *Computer Networks*, vol. 151, pp. 211–223, 2019.

[19] Liangrui Tang, Zhilin Lu and Bing Fan, "Energy efficient and reliable routing algorithm for wireless sensors networks," *Applied Sciences*, vol. 10, no. 5, pp. 1885–1889, 2020.

[20] Thar Baker, Jose M. García-Campos, Daniel Gutiérrez Reina, Sergio Toral, Hissam Tawfik, Dhiya Al-Jumeily and Abir Hussain, "GreeAODV: An energy efficient routing protocol for vehicular ad hoc networks," in *International Conference on Intelligent Computing*, pp. 670–681, Springer, Cham, 2018.

[21] Saijun Shao, Gangyan Xu and Ming Li, "The design of an IoT-based route optimization system: A smart product-service system (SPSS) approach," *Advanced Engineering Informatics*, vol. 42, p. 101006, 2019.

[22] Piyush Agarwal, Priya Matta and Sachin Sharma, "Analysis based traffic flow control decision using IoT sensors," *Materials Today: Proceedings*, vol. 46, no. 20, pp. 10707–10711, 2021.

[23] Amit Juyal, Sachin Sharma and Priya Matta, "Deep learning methods for object detection in autonomous vehicles," in *2021 5th International Conference on Trends in Electronics and Informatics (ICOEI)*, pp. 751–755, Tamilnadu, 2021.

[24] Zongshan Wang, Hongwei Ding, Bo Li, Liyong Bao and Zhijun Yang, "An energy efficient routing protocol based on improved artificial bee colony algorithm for wireless sensor networks," *IEEE Access*, vol. 8, pp. 133577–133596, 2020.

[25] Mukhdeep Singh Manshahia, "Grey wolf algorithm based energy-efficient data transmission in Internet of Things," *Procedia Computer Science*, vol. 160, pp. 604–609, 2019.

[26] Sakshi Painuly, Sachin Sharma and Priya Matta, "Future trends and challenges in next generation smart application of 5G-IoT," in *5th IEEE International Conference on Computing Methodologies and Communication (ICCMC 2021)*, Tamilnadu, 2021.

[27] Priya Kohli, Sachin Sharma and Priya Matta, "Security of cloud-based vehicular ad-hoc communication networks, challenges and solutions," in *IEEE 6th Edition of the International Conference on Wireless Communications Signal Processing and Networking Wispnet*, Tamilnadu, 2021.

[28] S. U. Maheswari, N. Usha, E. M. Anita and K. R. Devi, "A novel robust routing protocol RAEED to avoid DoS attacks in WSN," in *Proceedings of the IEEE 2016 International Conference on Information Communication and Embedded Systems (ICICES)*, pp. 1–5, Chennai, India, 25–26 February 2016.

[29] Piyush Agarwal, Priya Matta and Sachin Sharma, "Comparative Study of Emerging Internet-of-Things in traffic management system," in *2021 5th International Conference on Trends in Electronics and Informatics (ICOEI)*, pp. 422–428, Tamilnadu, 2021.

# CHAPTER 9

# Sustainable Energy Solutions

## *Integration with Renewable Energy Sources*

Kimmi Verma, Deepti Agarwal

## CONTENTS

## 9.1 INTRODUCTION

Renewable energy sources bring a lot of positive aspects with respect to the environment. However, any change in the conventional system comes

DOI: 10.1201/9781003218760-9

with its own challenges. In addition to manufacturing and disposal-related issues, operational issues related to grid connectivity and grid frequency control pose a great challenge.

Presently the contribution of renewables (solar and wind being the only major contributors) is 68 GW, or 18.8% of the total 364-GW installed capacity. The share of renewables is low as compared to conventional thermal power generation. However, per the government of India's plan, by 2021–22, the installed renewable capacity will rise to 175 GW, with a total installed capacity reaching 480 GW [1]. This means that the share of renewables will rise to 36.5% [2].

This additional power generation for a limited time of the day is bound to create grid disturbances due to demand and supply mismatch. This issue can be addressed by two methods.

The first method is by increasing the inertia of the grid, that is, creating an infrastructure that can store energy when the power generation is more than the demand and release energy when the demand is more than the generation.

This inertia can be increased by adding large battery storage to the grid and/or by providing pumped hydropower storage with reversible pump turbine sets. However, this solution requires large investments and may take some time to become realizable.

Until such a time, the second method is to be adopted, that is, loading and unloading of the existing power-producing units. When the sun is shining, conventional power plants will be required to shut down or operate at very low loads, whereas during the night, when the share of renewables drops drastically, conventional plants will be required to make up for the reduction in renewable power generation. It is important to note that operating a power-generating unit at very low loads is a challenge and is not efficient operation. Also, there will be some uncertainty such as cloudy weather, high wind speed, and so on, when even in the daytime, conventional power plants will be required to increase their load for short durations as long as such a situation persists. A typical 500-MW coal-based thermal power unit takes 6 to 7 hours to start up from a cold condition. With addition of renewables to the grid, such a long time to come up to grid demand levels will not be permitted. This leaves power operators with no choice but to keep units running at low loads, leading to inefficient operation. This will result in more emissions per unit of electricity generation, offsetting the benefits of green energy generation by renewables.

The entire power system is required to be designed/modified to ensure integration of renewables with following aspects:

- Grid security

- Reliability of generating units

- Minimize the cost of flexible operation

- Maximize renewable integration

- Minimize investments by optimal utilization of existing assets and infrastructure

## 9.2 BASIC DEFINITIONS RELATED TO FLEXIBLE OPERATION

### 9.2.1 Flexibility

The ability of a power plant to operate at varying loads per grid requirements without compromising the safety of the equipment and personnel.

### 9.2.2 Renewable Energy Sources

In the present context, solar and wind power generation are considered renewable energy sources (RESs), as other sources contribute relatively low power generation to the grid.

### 9.2.3 Minimum Load

The minimum load of a power plant unit is the minimum stable load at which the unit can operate safely for long duration without consuming significant equipment life.

#### 9.2.3.1 Startup Time

Startup time is defined as the time required to load the unit from a stand-still condition to a minimum load. This time also depends on the last operation time; that is, if the unit was in service a few hours before the start, it is called a hot start, and the time required for startup will be less. If the unit has been in shutdown conditions for few days or more before the start, it is called a cold start, and more time will be required for startup.

### 9.2.4 Ramp Rates

Ramp rates define the capability of the unit in terms of how fast the unit can load or unload from a minimum load to a full or any other load. Higher ramp rates indicate that the unit can reach the desired load faster.

## 9.3 FLEXIBLE OPERATION

To understand the requirements of flexible operation, it is required to understand the variation of power generation from various sources [3]; see Figure 9.1.

From the figure, it is understood that generation starts rising at 8:00 hrs and remains steady between 11:00 and 16:00 hrs. From there it starts reducing and remains low between 19:00 to 7:00 hrs the next day, and this cycle is repeated daily.

Presently, hydro power is balancing the variation in renewable generation. However, with high renewable power, the variation will also be greater, and hydro power may not be able to play the balancing role. The onus of balancing the grid will fall on thermal power plants (mainly coal-based plants). Figure 9.2 presents the demand variation throughout the day and the contribution from different fuel sources for a critical day in 2021.

From Figure 9.2, it is evident that the power demand throughout the day varies from 171 to 190 GW, that is, a variation of 10%, which is reasonably stable. However, the renewable generation varies from 22 to 108 GW, a variation of 490%. Accordingly, thermal power generation varies from 33 to 117 GW, a variation of 355%.

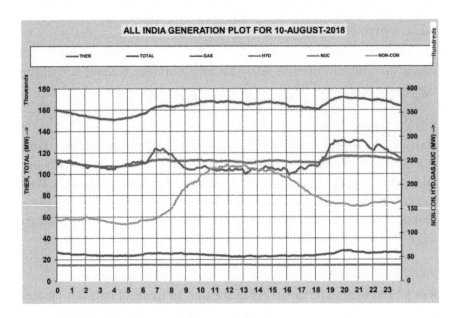

FIGURE 9.1   Power generation data for a typical day

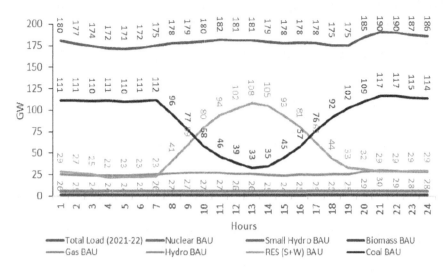

FIGURE 9.2    Load demand and contribution from different fuel sources in meeting demand—2021

This indicates that the thermal units, which are operating at 90% capacity during the peak generation of the day, may have to back down to ~30% load on an average if all these units are to be kept in service. Some units may not operate at such low loads and may limit their load to a higher value such as 40%. This will call for further low load operation of the remaining units or complete shutdown of the units. It should be noted that each shutdown costs a significant amount of money and is not recommended. Hence, many power plant units may be required to operate between 20% and 25%.

Another aspect that plays a significant role is the ramp up and ramp down. From 07:00 AM to 01:00 PM, thermal generation reduces from 112 to 33 GW. This translates into an average ramp-down rate of ~220 MW/min. Similarly, from 01:00 PM to 09:00 PM, thermal generation reduces from 33 to 117 GW. This translates into an average ramp-up rate of ~175 MW/min.

Not all plants can ramp up or down simultaneously or at the same rate. Some plants may be required to ramp up or down faster to support grid stability. A ramp rate of 0.5% per minute is required, 1% per minute is desirable, and 3% per minute is preferred. However, high ramp rates adversely affect the unit's stable operation, and it may trip and cause severe grid disturbances.

### 9.3.1 Options for Flexible Power

Presently, hydro power plants perform grid-side demand and supply balancing. However, with large integration of renewable power sources, additional methods are required to perform balancing. The following are the main methods being considered.

### 9.3.2 Pumped Hydropower Storage

Pumped hydropower storage (PHS) is a modified hydro power plant that can operate as a hydro turbine as well as a pump. When power generation is higher, such as during the daytime, PHS will increase the grid demand and pump water from lower elevations to higher elevations. On the other hand, when generation is lower and demand is more, such as during the night, PHS will operate as a hydro turbine and support the grid to meet the demand.

In this way, PHS can play a significant role in flexibility. However, presently there is a small capacity of 4,785 MW PHS installed.

### 9.3.3 Battery Storage

Similar to PHS systems, battery storage is being considered as a potential grid-side demand and supply balancing. When power generation is higher, such as during the daytime, the battery storage systems will increase the grid demand and charge the batteries. On the other hand, when the generation is lower and demand is higher, such as during the night, battery storage will operate as a power source and support the grid to meet the demand.

The cost of batteries has decreased significantly; hence, battery storage may offer a good solution for grid flexibility.

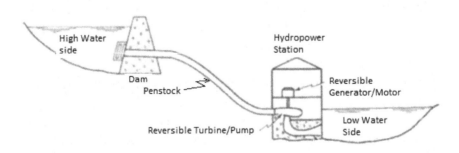

FIGURE 9.3   Schematic of pumped hydropower storage system

9.3.4 Flexibilization of Thermal Power Plants
Flexible operation modification/design should:

- Minimize the cost of flexible operation

- Minimize investments by optimal utilization of existing assets and infrastructure

In view of these objectives, investment in pumped hydropower storage or battery storage may be planned over the medium to long term; large investment is not foreseen in the near future in this area. Hence, until pumped hydropower storage or battery storage begins to play a major role, alternative solutions are required, and operation of thermal power plants suiting the requirements of the grid presents the most cost-effective solution.

Flexible operation requires two main capabilities: stable operation and low loads (40% or below) and fast ramp up or ramp down suiting the grid demand. It is more economical to operate the plant at low loads than to shut down the whole plant, as there is a significant cost involved with each startup. A standard thermal power plant operates with coal, but for startup purposes, oil firing is required, which is significantly costlier than coal.

There is no generic concept or single implementation plan for power plants, as each plant has its own specifics, technology requirements, and site conditions. However, some generic solutions and their limitations are as follows.

*9.3.4.1 Low Load Operation*
Low load operation is carried out for units when the unit is starting or shutting down. However, these operations are transient in nature, and unit load is constant/stable at any point of operation. Further, during such conditions, the fuel is generally oil, making the operation of the unit uneconomical.

For low load operation, the unit is brought to a partial load at which the unit has operated satisfactorily in the past. Generally, this load value is around 60% to 70% of the unit load. In some cases, this load may be as low as 55%.

The unit load is then gradually reduced by 5%, say, to 50% load, and the unit is allowed to stabilize for some duration (15–30 min). During this time, steam temperatures are observed, as these parameters play an important role in unit operation.

The unit load is then further reduced by 5% to 45%, and the unit is allowed to stabilize for some duration (15–30 min). During this time, besides steam temperatures, boiler burner flame also becomes critical. The reason for this is that with a reduction in loads, coal flow reduces, but the air flow does not reduce proportionately, making the coal–air mixture lean. If the air flow is also reduced, the possibility of flame out increases. Steam temperatures continue to play a critical role in unit operation. It is important to note that high steam temperatures may damage the boiler tubes, boiler-turbine interconnecting piping, and turbine components, whereas low-temperature steam may damage the steam turbine blades due to moisture formation inside the turbine.

When the unit load is further reduced by 5% to 40%, all parameters are reviewed more critically. If any abnormality is observed in the parameters, the unit load is increased to avoid tripping. If the parameters are found to be within an acceptable range, the unit load can be further reduced with a step load change of 2% in place of 5%.

There are various factors that play a critical role in the low load capability of the units:

1 Coal quality

2 Unit ageing

3 Operation and maintenance

4 Control systems

Some units have demonstrated a low load capability up to 30% load for short duration. However, operation at 40% load has been considered a benchmark for low load operation.

Typical variation in the operating parameters with a variation in load for a 250 MW unit is estimated using simulation software, indicated in Figures 9.4 to 9.6. It can be seen from the simulated data that there is a significant variation in the steam parameters.

Besides variation in the parameters, it is important to note that such variation plays a major role in equipment life consumption. Fluctuations in the pressure and temperature values may lead to fatigue failure of the component, and studies are presently underway to estimate this.

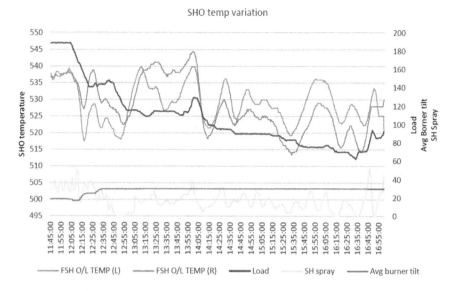

FIGURE 9.4  Super heater outlet temperature variation with respect to load over a period of time

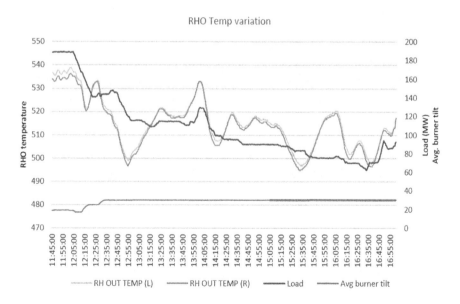

FIGURE 9.5  Reheater outlet temperature variation with respect to load over a period of time

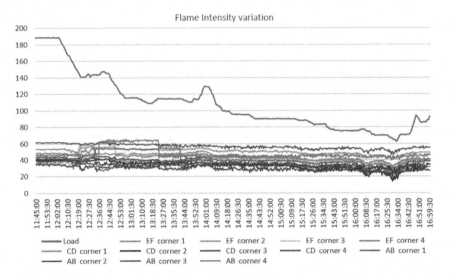

FIGURE 9.6  Flame intensity variation with respect to load over a period of time

### 9.3.5 Faster Ramp Up and Ramp Down

As the capability for low load operation of thermal power plants is established, to meet grid demand, the next challenge is to vary the load per grid requirements.

To meet the requirements, units' ramp-up and ramp-down capabilities play the most important role. All units are required to ramp up or ramp down sometimes, such as during startup, shutdown, or during equipment failures. However, the rate of ramp up or ramp down is critical for flexible operation to meet the grid requirements.

If there is no restriction, units should be ramped up at a very slow rate (0.5% or less) so that all the parameters remain within design limits. With higher ramp-up rates, the parameters deviate from the design condition. Some deviation from the design parameters is allowed; however, high ramp rates lead to higher deviations and may lead to unit tripping, higher equipment life consumption, and so on.

Typical variation in the parameters with unit ramp up and ramp down at 3% rates for a 250-MW unit is estimated using simulation software, indicated in Figures 9.7 to 9.9.

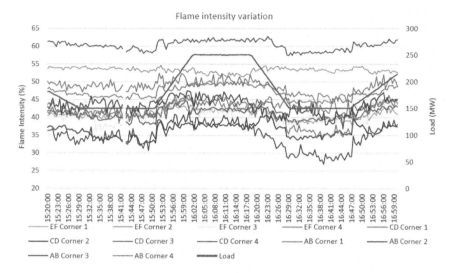

FIGURE 9.7    Flame intensity variation with respect to load and ramp rates over a period of time

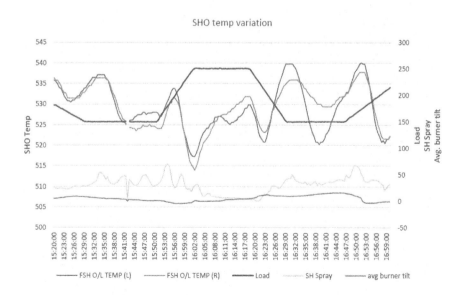

FIGURE 9.8    Super heater outlet temperature variation with respect to load and ramp rates over a period of time

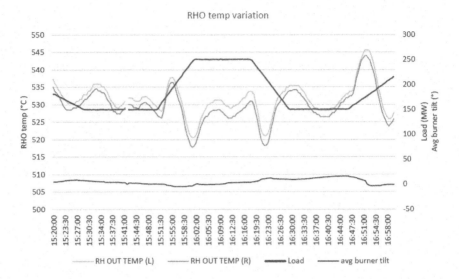

FIGURE 9.9  Reheater outlet temperature variation with respect to load and ramp rates over a period of time

## 9.4  CONCLUSION

Renewable power is required for the sustainable development of the country. It will also play a major role in limiting the effect of global warming. The government of India, under the Paris Agreement, has already committed to a reduction in greenhouse gas emissions. Hence, the challenges related to renewable power integration with the grid are to be addressed in the near future. There are some solutions available, such as pumped hydropower storage and battery storage. However, these solution are cost intensive and are not preferred for our country.

In the world of automation, a smart power plant can be proposed with the help of IoT, where sensors can be implanted to sense or estimate weather conditions. Accordingly, power generation can be switched from renewable to nonrenewable sources.

Flexible operation of thermal power plants offers a cost-effective solution for grid load requirements for integration of renewable power. However, fluctuations in parameters are required to be controlled by adopting advanced control systems. This will also help in reduction of the lifetime of the equipment, and plant will operate for a relatively longer duration.

## REFERENCES

[1] "Flexible Operation of Thermal Power Plant for Integration of Renewable Generation – A Roadmap for Flexible Operation of Thermal, Gas and Hydro Power Stations to Facilitate Integration of Renewable Generation," – Central Electricity Authority, Ministry of Power, Government of India, January 2019.

[2] Ministry of Power, Government of India, [Internet]. 2019. www. https://powermin.nic.in. Accessed 7 December 2019.

[3] Power System Operation Cooperation Ltd. (POSOCO), "National Load Dispatch Centre," *Monthly Report*, August 2018.

# CHAPTER 10

# Recent Advances and Future Trends of IoT-Based Devices

Punit Kumar Singh, Sudhakar Singh, Ashish, Hassan Usman, Shabana Urooj

## CONTENTS

DOI: 10.1201/9781003218760-10

## 10.1 INTRODUCTION

The Internet of Things (IoT) is gradually becoming an essential element of our lives, and it can be felt all around us. Overall, the IoT is an invention that brings together a wide range of smart arrangements, frameworks, intelligent devices, and sensors. The Internet of Things is a network of connected devices, computers, and digital devices that transport data using unique identifiers. Smartphones, laptops, wearables, and sensors are all connected to the Internet of Things and share data. The way patients receive healthcare will never be the same again. Every day, new breakthroughs in this sector are made. The Internet of Things in the healthcare system is a new paradigm that permits connection between medical equipment and sensors over the internet to ease medical personnel's tracking and remote monitoring of patient health status. Beyond that Internet of Things in healthcare systems also includes patients with heart monitor implants, automatic internal deliberators, and ventricular assist device implants, for which the doctor can collect data on performance and condition without removing them. The Internet of Things is a system of inter-related devices connected to the internet to transfer and receive data from one device to another without human intervention. In the phrase "Internet of Things," the word "Things" could be any medical device or instrument that has a sensor/transducer that detects any changes or information and transfers the data to another device automatically via the Internet. The Internet of Things has enabled services that do not require human interaction in a variety of applications, including continuous remote patient monitoring (RPM). The evolution of the Internet of Things in health care systems is quite potentially lifesaving, because by collecting data from bedside devices and viewing patient information and diagnoses in real time from distant locations, the entire healthcare system could be improved. The Internet of Things in healthcare will drastically reduce the cost of services; improve the ease of tasks for doctors, nurses, biomedical engineers, and other medical staff in the healthcare system; and will also remove burdens on caregivers. The impact of the Internet of Things can be seen in the remote monitoring of babies' environmental factors in incubators, as well as the remote monitoring of patients at risk of myocardial infarction, hypertension, and other heart-related diseases that necessitate constant monitoring, diagnosis, and early treatment. The Internet of Things might be used to assess the operation of medical devices over time to assist clinical engineers in evaluating equipment conditions. It is generally recognized that the Internet is a major source of security threats, and cyber-attacks have unlocked many doors for hackers, making data and information vulnerable. Nonetheless, the IoT is focused on giving the best accessible answers to adapt to information and data security concerns. Security will soon be the main issue of IoT in medical care. Therefore, the development of a protected pathway for coordinated efforts between interpersonal organizations and security issues is a hot issue in the IoT in the medical care framework, and IoT designers are working diligently on this. Clinical hardware with CPUs, sensors, IoT doors, microcontrollers, and UIs like PDAs are essential for the IoT framework in medical care.

The Internet of Things is gradually becoming an essential element of our lives, and it can be felt all around us. Overall, the Internet of Things is an innovation that associates an assortment of smart frameworks, systems, devices, and sensors. TeleHealth is an FDA-approved, HIPAA-compliant platform that links patients with a nationwide network of licenced doctors via the Internet, Internet of Things, video chats, cellphones, and electronic medical record (EMR) clouds 24 hours a day, seven days a week.

The following are the foremost types of IoT services: first, smart wearable devices may be used by patients who need to gather data about their health state, such as heart rate, blood pressure, and glucose level, using sensors on wearable technology and sending it to smartphones. Patients' health state can be checked at the same time. Second, the IoT can improve smart homes. When sensors can detect temperature changes, air-conditioning

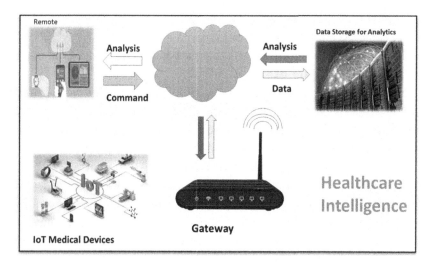

FIGURE 10.1   A portion of the potential application spaces of the IoT

systems can be monitored. Home security cameras may detect intruders and provide alerts to homeowners through mobile applications. Third, the IoT can monitor traffic and transportation systems in order to create smart cities. Data may be gathered and analyzed to better understand how traffic networks and transportation systems are evolving.

The remainder of the chapter is structured as follows: The "Literature Review" section will give the most recent information on significant research that addresses different difficulties and concerns in the IoT. "The Internet of Things Architectural Design" goes over IoT blocks, functions, and architecture in depth. The section "The Main Issues, Challenges, and Concerns of the IoT" discusses crucial essential concerns and challenges of the IoT. The section "Major IoT Applications" describes developing IoT application fields. The significance and relevance of big data and its investigation are described in the section "Importance of Big Data Analytics in IoT." Finally, the chapter finishes with the "Conclusions" section.

## 10.2 LITERATURE REVIEW

The IoT offers a multidisciplinary platform that will help a variety of sectors, including the environment, industry, public/private sectors, medical, transportation, and so on. Various academics have discussed the Internet of Things in various ways based on their distinct interests and characteristics. The Internet of Things has emerged in many sectors of the healthcare system, as illustrated in Figure 10.2, such as in telemedicine, smart ambulances, wearable devices, robotic surgery, heart monitor implants, traffic management, automatic defibrillators, and smart cities. A few critical IoT drivers have recently dealt with business.

The most significant IoT drivers that have held onto most of the market are portrayed in Figure 10.3.

These IoT projects are conveyed universally across the Americas, Europe, and the Asia/Pacific region. It very well may be apparent that the American landmass offers more to medical services and smart storage network drives, while the European mainland offers more to smart city projects.

Figure 10.4 portrays worldwide IoT projects [1]. In contrast with others, it is obvious that businesses, smart cities, smart energy, and smart vehicle-based IoT projects rule the market. Smart urban communities are perhaps the most famous IoT application region, and they incorporate smart homes, too. A smart house is composed of IoT-empowered home machines, a cooling/warming framework, a TV, real-time sound/

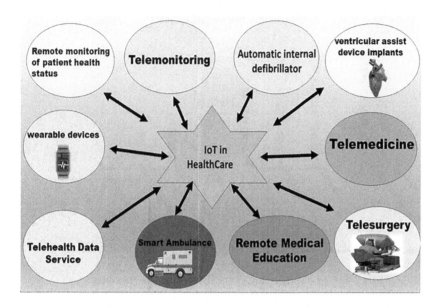

FIGURE 10.2    IoT applications in healthcare systems

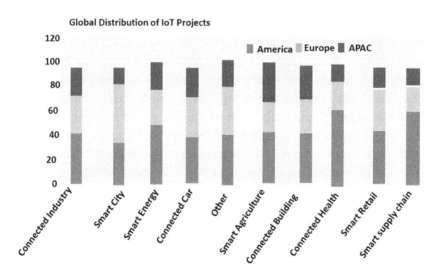

FIGURE 10.3    The worldwide circulation of IoT projects among America (the United States, South America, and Canada), Europe, and the Asia and Pacific region

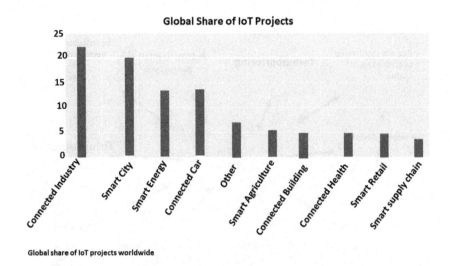

**Global Share of IoT Projects**

Global share of IoT projects worldwide

FIGURE 10.4  The global distribution of IoT projects sector wise

video devices, and security frameworks that speak with each other to give the ideal solace, security, and energy utilization. All of this communication is carried out over the Internet by an IoT-based dominant controller unit. The notion of smart cities has grown in popularity over the previous decade, attracting a large amount of research activity [2]. By 2022, the smart home business economy will have surpassed the $100 billion mark [3]. Smart homes not only give in-house comfort but also help homeowners save money in a variety of ways, such as lowering energy usage, which results in a cheaper power bill. Beside smart houses, another region that falls under the umbrella of smart city is smart automobiles. Current cars are equipped with complex contraptions and sensors that deal with most of the segments, from the headlights to the motor [4]. The Internet of Things is devoted to the advancement of new smart vehicle frameworks that coordinate remote availability from vehicle to vehicle and vehicle to driver to empower prescient support while likewise conveying an agreeable and safe experience of driving [5].

[3] gives a review of IoT advances for energy board control to profit smart city applications. The extent of the IoT is very expansive, and the IoT will, for all intents and purposes, cover all application regions sooner rather than later. They expressed that energy preservation is a fundamental part of human progress and that the IoT may help in the advancement of a smart energy control framework that would set aside both energy and

cash. They portrayed Internet of Things engineering in association with the idea of a smart city. The creators likewise noted that the novelty of IoT equipment and programming is quite possibly the most difficult factor in doing this. They suggested that these issues be settled to offer an IoT framework that is reliable, effective, and easy to understand. [6] resolved the issue of urbanization in urban areas. The movement of people from rustic to metropolitan districts has brought about an rise in city populations. Therefore, answers for smart versatility, energy, medical services, and framework are required. Quite possibly the main application region for IoT designers is smart urban areas. It includes various issues, for example, traffic, air quality administration, local area security arrangements, smart stopping, smart lighting, and smart trash assortment. They expressed that the IoT is striving to address these difficulties. With expanding urbanization, the interest in improved smart city frameworks has set opened doors for business visionaries in the field of smart city innovation. The creators arrived at the resolution that IoT-empowered innovation is basic for the production of manageable smart urban communities. Another basic part of IoT that needs consideration and broad examination is security and protection. Weber [7] underlined these issues and suggested that, as an additional benefit, a private partnership embracing the IoT should join information confirmation, access control, resistance to attacks, and customer security in their business exercises.

Weber said that to characterize overall security and protection issues, IoT designers should think about the geological limitations of different countries. A conventional design should be made to meet overall protection and security needs. Prior to building a completely useful IoT system, it is emphatically encouraged to investigate and understand the hardships and hindrances in protection and security. Afterward, [8] found a safety blemish in an IP-based IoT framework. They expressed that the web acts as the spine for the device network in an IoT framework. Thus, security worries in IP-based IoT frameworks are a significant concern. This additionally includes the cooperation of a reliable outsider and security norms. It is profoundly alluring to have a security design that can deal with a limited to huge scope of things in the web of things. Also, security configuration should consider the existence cycle and abilities of each object in the IoT framework. Per the examination, the Internet of Things has led to another kind of correspondence among assorted things all through the organization, and conventional start-to-finish web conventions can't offer the necessary help for this correspondence.

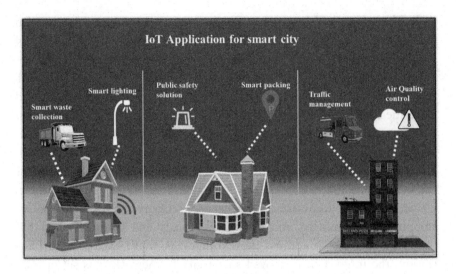

FIGURE 10.5    Internet of Things for smart cities

Thus, new conventions should be worked on in light of these inter-
pretations at entryways to guarantee start-to-finish security. Also, each
layer answerable to correspondence has its own arrangement of safety
concerns and needs. Subsequently, meeting the standards for one layer
would leave the framework helpless, and security ought to be guaran-
teed for all layers. Another test of the IoT that requires possible answers
for upgraded security is verification and access control. [9] proposed a
framework for confirmation and access control. Confirmation is basic to
validate participating parties and ensure the security of secret information. [9] proposed a validation procedure dependent on the Elliptic Arc
cryptosystem and tried it against diverse security dangers, for example,
snooping, man-in-the-middle attack, key control, and replay attack.
They expressed that their recommended frameworks can empower further developed validation and access control in IoT-based communication. [10] later introduced a two-way verification framework for the
IoT dependent on datagram transport layer security (DTLS). Assailants
on the web are consistently dynamic in their endeavors to take secure
data. In an IoT-based correspondence organization, the recommended
arrangement may offer message security, trustworthiness, realness,
and secrecy, just like memory overhead and start-to-finish inactivity.
[11] introduced a unique way to deal with cloud stages for information-
driven IoT applications.

A prerequisite for a satisfactory device, programming arrangement, and framework is productive answers to serve countless IoT applications working in the cloud. IoT specialists and designers are continually associated with building arrangements that consider both enormous platforms and assorted IoT things and devices [12]. The software defined networks-based security engineering is more versatile and proficient for IoT. The key job of a protected sensor organization semantic sensor network (SSN), as stated by [13], is to offer information security, guard from repetition attacks, etc. They examined two notable SSN administrations: TinySec [14] and ZigBee [15]. They noticed that, while both SSN frameworks are effective and dependable, ZigBee gives higher security but uses more energy, while TinySec requires less.

In further research, another compositional plan was proposed [22], MiniSec, to give strong security while utilizing less energy and exhibited its viability on the Telos platform. Trust is a critical issue in IoT, as indicated by [22]. Clients can understand and believe IoT administrations and applications because of trust, which takes out vulnerability and dangers [30]. They investigated a few difficulties in trust and addressed their importance comparable to IoT engineers and customers. Per [26], interoperability is significant in the IoT in light of the fact that it empowers the incorporation of devices and administration from numerous heterogeneous platforms to convey a productive and reliable arrangement. A few extra investigations [28, 39] underlined the viability of interoperability and investigated the different issues that interoperability faces in the IoT. [40] investigated an Earth-wide temperature boost by proposing an IoT-based natural reconnaissance framework. Existing methods, they said, were time concentrated and required significant human communication. Furthermore, a standard visit is important to assemble information from sensors at the area under assessment. Also, certain data was forgotten about, resulting in a less-than-precise examination. Accordingly, an IoT-based structure may deal with this test and give exact outcomes in examination and expectation.

[41] expressed their concern for private waste water treatment. They found different problems in surplus water management measures and the observing framework, and they proposed suitable cures dependent on the IoT. According to them, the IoT can be exceptionally useful in waste water treatment and cycle checking. Farming is a significant area from one side of the planet to the other. Agribusiness is impacted by an assortment of components like geology, nature, and so on. As indicated by [33],

TABLE 10.1  Investigation Correlation on Explicit Appraisal Factors

| Research | Major Research Directions | Based on the Assessment Criteria, a Comparison Is Made | | | | |
|---|---|---|---|---|---|---|
| | | RT | DL | AV | CT | EC |
| Zhou et al. [16] | Protection and confidentiality | — | x | — | — | X |
| Sfar et al. [17] | Design, protection, and confidentiality | x | — | x | x | — |
| Gaona-Garcia et al. [18] | Design, protection, and confidentiality | — | — | x | — | X |
| Behrendt [19] | Smart city, transport, and healthcare | — | x | — | x | X |
| Zanella et al. [20] | Smart city, transport, and healthcare | x | — | x | — | — |
| Khajenasiri et al. | Environment, power and energy, smart city | x | — | x | x | X |
| Alavi et al. [3, 6] | Transport and healthcare | x | — | x | — | — |
| Weber [7] | Protection and confidentiality | x | x | — | — | X |
| Heer et al. [8] | Protection and confidentiality | x | — | — | — | X |
| Liu et al. [9] | Verification and identification | x | x | — | x | — |
| Kothmayr et al. [10] | Protection and confidentiality | — | x | — | — | — |
| Li et al. [11] | Protection and confidentiality, management and control | — | x | x | x | — |
| Luk et al. [13] | Protection and confidentiality, architecture | x | x | — | — | — |
| Sebastian and Ray [21] | Smart city, transport and healthcare, architecture | x | — | x | — | X |
| Yan et al. [22] | Verification and identification, quality of service (QoS) | x | — | — | — | X |
| Dierks and Allen [23] | Standardization | x | — | x | — | — |
| Pei et al. [24] | Standardization, verification, and identification | — | x | — | — | X |
| Roman et al. [25] | Protection and confidentiality | x | x | — | — | — |
| Noura et al. [26] | Interoperability | — | x | x | — | — |
| Palattella et al. [27] | Interoperability, dependability, scalability | x | — | x | — | X |

| Reference | Factors | RT | DL | AV | CT | EC |
|---|---|---|---|---|---|---|
| Yan et al. [22] | QoS, management and control, verification and identification | — | x | — | x | — |
| Pereira and Aguiar [28] | Interoperability, QoS, scalability, identification | x | — | x | — | — |
| Pereira and Aguiar [28] | Interoperability, QoS, scalability | x | — | x | — | X |
| Clausen et al. [29] | Data processing, reliability | — | x | x | x | — |
| Bao et al. [30] | Scalability, security and privacy | x | x | x | — | — |
| Li et al. [31] | Protection and confidentiality, dependability | — | x | — | — | — |
| Zhang [32] | Protection and confidentiality, data processing | — | x | x | — | — |
| Qiu et al. [33] | Agriculture, environmental | x | x | — | — | X |
| Fang et al. [34] | Environmental | x | x | x | — | — |
| Montori et al. [35] | Interoperability, dependability | x | x | x | — | X |
| Distefano et al. [36] | Interoperability, scalability | x | — | — | — | X |
| Temglit et al. [37] | QoS, reliability | — | x | x | — | — |
| Talavera et al. [38] | Agriculture, industrial, environmental | x | — | — | x | X |

RT = reaction time, DL = dependability, AV = availability, CT = cost, and EC = energy usage are all factors to consider.

the innovation used to direct biological systems is emerging and has a low insight level. As indicated by them, it very well might be a potential application region for IoT designers and scientists. [33] fostered a smart checking platform for engineering dependent on the IoT for office ranch environments, composed of four-layer techniques to control horticulture biological systems. Each layer is accountable for a specific undertaking, and the engineering, all in all, is fit to create a superior environment with less human involvement.

Environmental change because of worldwide temperature alteration is another significant issue all through the world. To provide a powerful ecological observing and control framework, [34] introduced a coordinated data framework internet information services (IIS) that incorporates the IoT, geo-informatics, distributed computing, a worldwide situating framework (GPS), a topographical data framework (GIS), and e-science. The proposed IIS further develops information social occasion, preparing, and dynamic for environment control. Another major worldwide issue is air contamination. There is some hardware and a few techniques accessible for estimating and controlling air quality. [42] proposed Air Cloud, a cloud-based air quality and checking framework. Nature of service (QoS) was distinguished by [37] as a critical issue and a difficult cycle in the evaluation and determination of IoT devices, conventions, and administration. QoS is a basic rule for drawing in and acquiring buyer trust in IoT administration and devices. They contrived a dispersed QoS choice strategy. The spread constraint optimization issue and the multi-agent digm served as the foundation for this technique. Furthermore, the method was tested using a series of experiments in realistic distributed systems. Another critical feature of the IoT is its compatibility with environmental and agricultural regulations. In a survey, [38] concentrated on this topic and highlighted the essential functions of IoT for agro-industrial and environmental issues. They stated that IoT initiatives in these sectors are apparent. The Internet of Things is working on existing innovation and profiting ranchers and society.

In [43], the significance of IoT-based patient wellbeing checking was featured. They argued that IoT devices and sensors, in combination with the web, can help in the observation of patient wellbeing. They likewise offered a system and procedure for accomplishing their aim. Table 10.1 sums up the examinations and research headings, along with an investigation correlation on explicit appraisal factors.

## 10.3 INTERNET OF THINGS ARCHITECTURAL DESIGN

IoT engineering is composed of five significant layers that govern the entirety of the usefulness of IoT frameworks. These levels are the discernment layer, the organization layer, the middleware layer, the application layer, and the business layer. At the bottom of IoT design is the discernment layer, which is composed of actual devices like sensors, smart wearable devices, embedded patient heart screen, programmed inward defibrillator, RFID chips, standardized tags, and other actual things that are connected together in an IoT organization. These devices accumulate patient information to send it to the organization layer. The organization layer goes about as a conductor for data to be sent from the discernment layer to the data handling model. This information transmission may utilize any wired or remote means, including LAN 5G/4G/WDMA/3G, Wi-Fi, Bluetooth, and so on. The middleware layer is the powerful layer. The significant task of this layer is to deal with data from the organization layer and settle on choices dependent on the aftereffects of continuous monitoring. A business layer sits on top of the engineering, controlling the entire IoT framework, its applications, and administration. The data and information acquired from the application layer are seen by the business layer, which is then used to select future destinations and techniques. Moreover, IoT models might be custom fitted to explicit prerequisites and application areas [12–13, 44].

In addition to the layered design, an IoT framework comprises a few practical elements that empower different IoT exercises like detecting, validation and personality, control, and organization [21]. Figure 10.6 delineates the utilitarian segments of IoT engineering. A few basic practical elements are responsible for I/O tasks, association issues, handling, sound/video checking, and capacity. These useful pieces, when joined, establish a productive IoT framework, which is basic for most execution. In spite of the way that different reference plans have been given specialized prerequisites, they are still a long way from the standard design that is appropriate for a worldwide IoT [45]. Therefore, reasonable engineering that can satisfy the worldwide IoT needs is yet to be implemented. IoT connections assume a fundamental part in IoT correspondence since they empower availability between IoT workers and IoT devices for an assortment of utilizations. [46]. The significant planning issues for viable IoT engineering in a heterogeneous setting are versatility, seclusion, interoperability, and transparency The IoT configuration should be intended to fulfill the requests of cross-space communications, multisystem reconciliation with the capacity to perform straightforward and adaptable administration assignments, significant information investigation and capacity, and easy-to-understand applications. Moreover, the engineering ought to be equipped to increase usefulness and add knowledge and computerization to the framework's IoT devices.

Moreover, the developing volume of huge information created by IoT sensor and device availability represents another issue. To adapt to the great needs of streaming information in an IoT framework, a proficient plan is vital. Cloud and haze/edge figuring are the two main IoT framework plans that assist with the handling, observing, and investigation of significant measures of information in IoT frameworks. Sensors and actuators have an essential impact on the main phase of the design. This present reality includes the climate, people, creatures, mechanical devices, smart vehicles, and structures, in addition to other things. Sensors identify signs and information streams from these certifiable things and convert them into information that would then be able to be broken down. Furthermore, actuators have the ability to intervene in reality, such as controlling the temperature of a room, slowing down a vehicle's speed, turning off music and lights, and so on. As a result, stage 1 aids in the collection of data from the actual world that may be valuable for future research. Stage 2 is in charge of collaborating with instruments and actuators, as well as entries and data collecting structures. The vast amount of data created in stage 1 is gathered and optimized in an organized manner appropriate for processing in this stage. Once the vast quantity of data has been collected and organized, it is ready to proceed to stage 3, edge computing. Edge computing is characterized as a distributed open architecture that enables the usage of IoT technologies and huge processing capacity from various places across the world. It is a powerful method for data processing streaming, making it ideal for IoT systems. Edge computing technologies in stage 3 deal with enormous amounts of data and provide different capabilities such as visualization, data integration from other sources, analysis utilizing machine learning algorithms, and so on. The last stage entails a number of critical activities, including in-depth processing and analysis, as well as providing feedback to improve the system's overall precision and accuracy. All that will be done on a cloud or in a server farm at this stage. To deal with this huge amount of streaming information, large information structures like Hadoop and Spark might be used; also, AI approaches can be utilized to better model expectations, bringing about a more exact and reliable IoT framework to satisfy current interests.

## 10.4 MAIN ISSUES, CHALLENGES, AND CONCERNS OF IoT

Despite their enormous potential to transform the healthcare sector, emerging technologies face unique challenges in their actual application. The most common difficulties center around the development of massive data by a large number of devices connected to the system, as well as the possibility of cyberattacks and data breaches. The fusion of IoT-based frameworks into all pieces of human existence just as the various advancements associated with information moved between implanted devices complicated matters and brought about a large number of difficulties and barriers. In a sophisticated smart tech society, these concerns are also a difficulty for IoT developers. As technology advances, so do the difficulties and need for sophisticated IoT systems. As a result, IoT developers must anticipate new challenges and propose solutions to them.

### 10.4.1 Concerns about Security and Privacy

Security and privacy are two of the most significant and difficult concerns in the Internet of Things because of many threats, cyberattacks, dangers, and vulnerabilities [47]. Deficient approval and verification; unreliable programming, firmware, and web interfaces; and a lack of vehicle layer encryption are the components that hinder device-level protection [48]. Security and privacy concerns are critical factors for developing trust in IoT systems in a variety of ways [23]. To keep away from security dangers and attacks, safety efforts should be joined at each level of IoT design [22]. To maintain the safety and isolation of IoT-based systems, numerous protocols are designed and efficiently applied on each tier of communication channels [24, 49]. secure socket layer (SSL) and datagram transport layer security (DTLS) are two cryptographic conventions that give security in different IoT frameworks by working between the vehicle and application layers [49] to ensure the security of IoT device communication.

Furthermore, if communication occurs within the IoT system utilizing wireless technologies, the structure becomes more exposed to safety concerns. As a result, certain approaches should be used to identify harmful acts as well as for self-healing or recovery. As a result, to establish communication between trustworthy parties, authorization and authentication must be maintained through a secure network [25]. Another difficulty is that different items communicating inside the IoT system have different privacy rules. As a result, before transferring data, everything in the IoT framework ought to have the option to approve the protection guidelines of other framework objects. Privacy, on the other hand, is a key consideration that allows consumers to feel secure and comfortable when utilizing IoT solutions.

## 10.5 INTEROPERABILITY AND STANDARDIZATION CONCERNS

Interoperability refers to the capacity to share data across different platforms and IoT devices. This information sharing is not dependent on the software and hardware that is in use. The issue of interoperability emerges as a result of the diverse nature of technologies and results utilized in IoT development. Technical, semantic, syntactic, and organizational interoperability are the four layers of interoperability [50]. IoT systems provide a variety of functions to promote interoperability, which facilitates communication between various items in a heterogeneous environment. Furthermore, multiple IoT platforms may be combined based on their features to give a variety of solutions for Internet of Things users [51]. Given the importance of interoperability, researchers have used many methods to deal with interoperability [52]. These solutions might be based on adapters/gateways, virtual networks/overlays, service-oriented architectures, and so on. Despite the fact that interoperability taking care of procedures mitigates a portion of the weight on IoT frameworks, there are still a few issues with interoperability that may be the subject of future examination [26].

## 10.6 LAW AND ORDER, MORALS, AND ADMINISTRATIVE RIGHTS

Another thought for IoT engineers is morals, enactment, and administrative rights. Numerous laws and guidelines are set up to protect the norm and moral principles and to hold people back from violating them. The primary contrast between morals and law is that morals are rules that people have faith in, and laws are restrictions forced by the public authority, although both morals and guidelines are expected to safeguard norms and quality and to hold people back from taking part in criminal conduct. Several real-world issues have been solved as a result of the development of the IoT, but it has also given birth to important ethical and legal difficulties [53]. Some of these difficulties are data security, protection, secrecy, trust, and data usability. Due to a lack of trust in IoT devices, the majority of IoT users favor government standards and laws pertaining to data protection, privacy, and safety. As a result, this issue must be addressed in order to preserve and enhance public trust in IoT devices and systems.

## 10.7 SCALABILITY, AVAILABILITY, AND DEPENDABILITY ARE ALL CRITICAL FACTORS TO CONSIDER

If a system can be expanded with new services, equipment, and devices without impacting performance, it is scalable. Supporting an immense number of devices with different memory, handling power, stockpiling limits, and transfer speeds is the significant issue with the IoT [28]. Another basic factor to consider is openness. Versatility and accessibility ought to be conveyed simultaneously in the IoT layered system. Cloud-based IoT frameworks are a smart illustration of adaptability since they offer satisfactory help for scaling the IoT network by adding more devices, stockpiling, and preparing power depending on the situation. This overall scattered IoT organization, then again, brings about another worldview for fostering a consistent IoT system that satisfies worldwide needs [54]. Another significant issue is the accessibility of assets to validate merchandise, paying little heed to area or season of interest. A few neighborhood IoT networks are associated in a dispersed manner with worldwide IoT platforms to exploit their assets and abilities. As a result, availability is a major problem [55]. Some services and resource availability may be disrupted due to multiple data transmission channels the use, such as satellite communication. As a result, an autonomous and dependable data communication route is necessary to ensure the continuous accessibility of resources and facilities.

## 10.8 SERVICE EXCELLENCE

Another important component of IoT is service quality. QoS is a tool for assessing the quality, proficiency, and execution of IoT devices, frameworks, and engineering [37]. Dependability, cost, energy feeding, security, accessibility, and facility time are the most significant mandatory QoS criteria for IoT applications [56]. The criteria of QoS standards must be met by a smarter IoT ecosystem. Furthermore, in order to assure the dependability of any IoT service or device, its QoS metrics must first be specified. Furthermore, consumers may be able to express their wants and expectations. There are several techniques that may be used to measure QoS; according to White et al. [57] A trade-off exists between quality variables and methods. As a result, in order to overcome this trade-off, high-quality models must be employed. Certain high-quality models, for example, ISO/IEC25010 [58] and OASIS-WSQM [59], are accessible in the literature and might be utilized to assess QoS calculation approaches. These models include an extensive variety of quality criteria that are more than enough for assessing QoS for IoT services. Table 10.1 summarizes the research on the IoT's main difficulties and concerns addressed previously.

## 10.9 MAJOR IoT APPLICATIONS

### 10.9.1 Smart Cities, Transportation, and Automobiles

With the concepts of smart cities, smart houses, smart automobiles, and smart transportation, the IoT is changing society's traditional civic framework into a high-tech structure. Fast advances in understanding the interest for and use of innovation at home are being accomplished with the assistance of supporting advances, for example, AI and natural language processing [60]. Numerous sorts of advances, for example, cloud worker innovation and remote sensor organizations, should be utilized related to IoT workers to create a successful smart city. Another basic thought is the smart city's natural effect. Therefore, energy utilization has expanded. Green and effective advances ought to likewise be thought of while planning and creating the smart city foundation. Moreover, modern devices being coordinated in new vehicles may recognize gridlock and prescribe an ideal backup course of action to the driver. This can help to lessen gridlock in the city.

### 10.9.2 Telemedicine

E-medicine refers to the remote delivery of healthcare services such as testing and consultations using telecommunications infrastructure. Telemedicine enables healthcare practitioners to examine, diagnose, and treat patients without needing to see them in person, providing expert advice to a patient at a remote place, or aiding a primary care physician in making a diagnosis. According to the American Medical Association (AMA), telemedicine could manage 78% of emergency treatment efficiently. Specialists and architects are attempting to make exceptionally productive IoT devices to screen an assortment of medical issues, including diabetes, obesity, and depression [61]. Several studies have taken into account a variety of healthcare-related problems. The term "telemedicine" refers to a system that uses telecommunications and information technology to diagnose patients and offer clinical healthcare from a distance.

Telemedicine services are in high demand. They save lives in crucial situations and emergencies. They include features such as live video streaming, chat boxes, automated prescription creation, and push notifications.

### 10.9.3 Telemonitoring

Personalized notifications that notify a patient's healthcare practitioner in situations of physical/mental trauma are also included in telemonitoring services. The use of information and communications technology to monitor and transmit data related to patient health status between geographically separated individuals [57] enables home monitoring of patients (living at home or in nursing or residential care homes) using external electronic devices in combination with a telecommunications system (using a land line phone or mobile telephone is referred to as telemedicine).

### 10.9.4 Telesurgery

Telesurgery is the use of telerobotic technology to allow a surgeon to perform an operation on a patient from a remote location: "surgery done on an inanimate trainer, animate model, or patient in which the surgeon is not present at the time the model or patient is operated on."

Haptic technology in telesurgery, which creates a virtual picture of a patient or incision, would allow a surgeon to see and feel what they are operating on. This technique is intended to provide a surgeon with the capacity to feel tendons and muscles as if they were in the patient's body.

### 10.9.5 Medical Education via the Internet

The IoT functionality may be utilized for distance learning, including self-directed modules and learning exercises that can be transformed to virtual

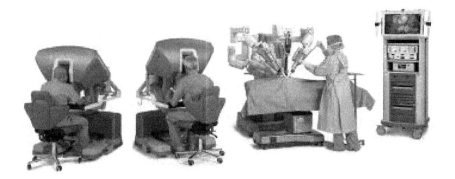

FIGURE 10.6    Telesurgery from Nate's surgery

interactions. As always, the resources are free to download and adapt to local conditions. In this situation, a virtual lab is an excellent example, where medical instruments that are highly expensive are easily accessed online.

### 10.9.6 The Significance of Data Analytics in the IoT in the Healthcare Sector

An IoT system is made up of a large number of interconnected devices and sensors. The number of these sensors and devices is continuously rising as the IoT network grows and expands. These devices connect with one another and send huge amounts of data via the Internet. According to system analysis, it is more important to incorporate cloud-based real-world applications in the future hospital management system. This small-scale model leads to an analysis to assess efficiency, accuracy, time-delayed sensor processing, networking and connection, and a comprehensive examination of hospital operations. The primary disadvantage is that it is more expensive to analyze sensor quality. It will be extremely beneficial for consumers if it is applied in the future to manage sensors as wearable devices and computers. Per analysis in the medical sector, IoT devices, particularly sensors used with patients, create large amounts of data on the patients' health every second. This huge quantity of information must be integrated into a single database and analyzed in real time in order to make rapid, accurate decisions, and big data technology is the ideal option for this purpose [62].

## 10.10 CONCLUSION

Recent developments in the IoT have piqued the interest of researchers and developers all around the world. Academicians and IoT programmers are also collaborating to expand the system on a wide scale and help society to the greatest extent feasible. There is a growing need to shift from clinic-centric to patient-centric healthcare. The IoT is projected to be a powerful enabler by allowing for the seamless connectivity of devices and cloud storage, as well as acting agents such as patients, hospitals, analytical laboratories, and emergency services. However, improvements are only feasible if we take into account the numerous difficulties and weaknesses in the current technical techniques. In this survey, various difficulties and obstacles that Internet of Things developers should consider while developing a better model are discussed. In addition, major IoT application areas where IoT developers and academics are involved are covered. The IoT not only provides services but also creates massive amounts of data. As a result, the significance of big data analytics is also highlighted, as it may give correct judgments that can be used to create a better IoT system.

## REFERENCES

[1] IoT application areas. https://iot-analytics.com/top-10-iot-project-application-areas-q3–2016. Accessed 5 April 2019.

[2] E. Park, A. P. Pobil and S. J. Kwon, "The role of Internet of Things (IoT) in smart cities: Technology roadmap-oriented approaches," *Sustainability*, vol. 10, p. 1388, 2018.

[3] K. Gatsis and G. J. Pappas, "Wireless control for the IoT: Power spectrum and security challenges," in *Proceedings of 2017 IEEE/ACM second international conference on internet-of-things design and implementation (IoTDI)*, pp. 18–21, Pittsburg, PA. INSPEC, April 2017. Accession Number: 16964293.

[4] Internet of Things. www.ti.com/technologies/internet-of-things/overview.html. Accessed 1 April 2019.

[5] T. Liu, R. Yuan and H. Chang, "Research on the Internet of Things in the automotive industry," *ICMeCG 2012 International Conference on Management of E-commerce and e-Government*, pp. 20–21, Beijing, China, October 2012.

[6] A. H. Alavi, P. Jiao, W. G. Buttlar and N. Lajnef, "Internet of Things-enabled smart cities: State-of-the-art and future trends," *Measurement*, vol. 129, pp. 589–606, 2018.

[7] R. H. Weber, "Internet of Things—New security and privacy challenges," *Computer Law and Security Review*, vol. 26, no. 1, pp. 23–30, 2010.

[8] T. Heer, O. Garcia-Morchon, R. Hummen, S. L. Keoh, S. S. Kumar and K. Wehrle, "Security challenges in the IP based Internet of Things," *Wireless Personal Communications*, vol. 61, no. 3, pp. 527–542, 2011.

[9] J. Liu, Y. Xiao and C. L. Philip-Chen, "Authentication and access control in the Internet of Things," in *Proceedings of 32nd International Conference on Distributed Computing Systems Workshops*, Macau, China, 2012. https://doi.org/10.1109/icdcsw.2012.23

[10] T. Kothmayr, C. Schmitt, W. Hu, M. Brunig and G. Carle, "DTLS based security and two-way authentication for the Internet of Things," *Ad Hoc Network*, vol. 11, no. 27, pp. 10–23, 2013.

[11] Y. Li et al., "IoT-CANE: A unified knowledge management system for data centric Internet of Things application systems," *Journal of Parallel and Distributed Computing*, vol. 131, pp. 161–172, 2019.

[12] F. Olivier, G. Carlos and N. Florent, "New security architecture for IoT network," in *Proceedings of International Workshop on Big Data and Data Mining Challenges on IoT and Pervasive Systems (BigD2M 2015), Procedia Computer Science*, vol. 52, pp. 1028–1033, 2015.

[13] M. Luk, G. Mezzour, A. Perrig and V. G. MiniSec, "A secure sensor network communication architecture," in *Proceedings of 6th International Symposium on Information Processing in Sensor Networks*, pp. 25–27, Cambridge, MA, April 2007.

[14] C. Karlof, N. Sastry and D. W. TinySec, "A link layer security architecture for wireless sensor networks," in *Proceedings of the Second ACM Conference on Embedded Networked Sensor Systems (SenSys 2004)*, MD USA, November 2004.

[15] ZigBee Alliance. ZigBee specification, "Technical report document" 053474r06, Version 1.0, ZigBee Alliance, June 2005.

[16] J. Zhou, Z. Cap, X. Dong and A. V. Vasilakos, "Security and privacy for cloud-based IoT: Challenges," *IEEE Communications Magazine*, vol. 55, no. 1, pp. 26–33, 2017. https://doi.org/10.109/MCOM.2017.1600363CM.

[17] A. R. Sfar, E. Natalizio, Y. Challal and Z. Chtourou, "A roadmap for security challenges in the Internet of Things," *Digital Communications and Networks*, vol. 4, no. 2, pp. 118–137, 2018, ISSN 2352-8648. https://doi.org/10.1016/j.dcan.2017.04.003.

[18] P. Gaona-Garcia, C. E. Montenegro-Marin, J. D. Prieto and Y. V. Nieto, "Analysis of security mechanisms based on clusters IoT environments," *International Journal of Interactive Multimedia and Artificial Intelligence*, vol. 4, no. 3, pp. 55–60, 2017.

[19] F. Behrendt, "Cycling the smart and sustainable city: Analyzing EC policy documents on Internet of Things, mobility and transport, and smart cities," *Sustainability*, vol. 11, no. 3, pp. 763–768, 2019.

[20] A. Zanella, N. Bui, A. Castellani, L. Vangelista and M. Zorgi, "Internet of Things for smart cities," *IEEE Internet of Things Journal*, vol. 1, no. 1, pp. 22–32, 2014.

[21] S. Sebastian and P. P. Ray, "Development of IoT invasive architecture for complying with health of home," in *Proceedings I3CS*, pp. 79–83, Shillong, 2015.

[22] Z. Yan, P. Zhang and A. V. Vasilakos, "A survey on trust management for Internet of Things," *Journal of Network and Computer Applications*, vol. 42, pp. 120–134, 2014.

[23] L. D. Xu, W. He and S. Li, "Internet of Things in industries: A survey," *IEEE Transactions on Industrial Informatics*, vol. 10, no. 4, pp. 2233–2243, 2014.

[24] M. Pei, N. Cook, M. Yoo, A. Atyeo and H. Tschofenig, "The open trust protocol (OTrP)," *IETF*, 2016. https://tools.ietf.org/html/draft-pei-opentrustprotocol-00.

[25] H. Van-der-Veer and A. Wiles, "Achieving technical, interoperability-the ETSI approach," ETSI White Paper No. 3, 2008. www.etsi.org/images/fles/ETSIWhitePapers/IOP%20whitepaper%20Edition%203%20fnal.pdf

[26] M. Noura, M. Atiquazzaman and M. Gaedke, "Interoperability in Internet of Things: Taxonomies and open challenges," *Mobile Networks and Applications*, vol. 24, no. 3, pp. 796–809, 2019.

[27] M. R. Palattella, M. Dohler, A. Grieco, G. Rizzo, J. Torsner, T. Engel and L. Ladid, "Internet of Things in the 5G era: Enablers, architecture and business models," *IEEE Journal on Selected Areas in Communications*, vol. 34, no. 3, pp. 510–527, 2016.

[28] C. Pereira and A. Aguiar, "Towards efficient mobile M2M communications: Survey and open challenges," *Sensors*, vol. 14, no. 10, pp. 19582–19608, 2014.

[29] H. Li, H. Wang, W. Yin, Y. Li, Y. Qian and F. Hu, "Development of remote monitoring system for henhouse based on IoT technology," *Future Internet*, vol. 7, no. 3, pp. 329–341, 2015.

[30] F. Jabeen, et al., "Adaptive and survivable trust management for Internet of Things systems," *IET Information Security*, vol. 15, no. 5, pp. 375–394, 2021. https://doi.org/10.1049/ise2.12029.

[31] L. Zhang, "An IoT system for environmental monitoring and protecting with heterogeneous communication networks," in *Proceedings 2011 6th International ICST Conference on Communications and Networking in China (CHINACOM)*, Harbin, China, 17–19 August 2011.

[32] F. Montori, L. Bedogni and L. Bononi, "A collaborative Internet of Things architecture for smart cities and environmental monitoring," *IEEE Internet Things Journal*, vol. 5, no. 2, pp. 592–605, 2018.

[33] T. Qiu, H. Xiao and P. Zhou, "Framework and case studies of intelligent monitoring platform in facility agriculture ecosystem," in *Proceedings. 2013 Second International Conference on Agro-Geoinformatics (Agro-Geoinformatics)*, Fairfax, VA, 12–16 August 2013.

[34] S. Fang et al., "An integrated system for regional environmental monitoring and management based on Internet of Things," *IEEE Industrial Informatics*, vol. 10, no. 2, pp. 1596–1605, 2014.

[35] S. Distefano, F. Longo and M. Scarpa, "QoS assessment of mobile crowd sensing services," *Journal of Grid Computing*, vol. 13, no. 4, pp. 629–650, 2015.

[36] N. Bizanis and F. A. Kuipers, "SDN and virtualization solutions for the Internet of Things: A survey," *IEEE Access*, vol. 4, pp. 5591–5606, 2016.

[37] N. Temglit, A. Chibani, K. Djouani and M. A. Nacer, "A distributed agent-based approach for optimal QoS selection in web of object choreography," *IEEE Systems Journal*, vol. 12, no. 2, pp. 1655–1666, 2018.

[38] J. M. Talavera et al., "Review of IoT applications in agro-industrial and environmental fields," *Computers and Electronics in Agriculture*, vol. 142, no. 7, pp. 283–297, 2017.

[39] A. Al-Fuqaha, M. Guizani, M. Mohammadi, M. Aledhari and M. Ayyash, "Internet of Things: A survey, on enabling technologies, protocols, and applications," *IEEE Communications Surveys & Tutorials*, pp. 2347–2376, June 2015.

[40] N. S. Kim, K. Lee and J. H. Ryu, "Study on IoT based wild vegetation community ecological monitoring system," in *Proceedings 2015 7th IEEE International Conference on Ubiquitous and Future Networks*, Sapporo, Japan, 7–10 July 2015.

[41] J. Y. Wang, Y. Cao, G. P. Yu and M. Yuan, "Research on applications of IoT in domestic waste treatment and disposal," in *Proceedings 11th World Congress on Intelligent Control and Automation*, Shenyang, China, 2014.

[42] Y. Cheng et al., "AirCloud: A cloud based air-quality monitoring system for everyone," in *Proceedings of the 12th ACM Conference on Embedded Network Sensor Systems*, pp. 251–265, Memphis, Tennessee, 3–6 November 2014.

[43] A. J. Jara, M. A. Zamora-Izquierdo and A. F. Skarmeta, "Interconnection framework for mHealth and remote monitoring based in the Internet of Things," *IEEE Journal on Selected Areas in Communications*, vol. 31, no. 9, pp. 47–65, 2013.

[44] J. Gubbi, R. Buyya, S. Marusic and M. Palaniswami, "Internet of Things (IoT): A vision, architectural elements, and future directions," *Future Generation Computer Systems*, vol. 29, no. 7, pp. 1645–1660, 2013.

[45] R. Nicolescu, M. Huth, P. Radanliev and D. D. Roure, "Mapping the values of IoT," *Journal of Information Technology*, vol. 33, no. 4, pp. 345–360, 2018.

[46] P. Hu, H. Ning, T. Qiu, Y. Xu, X. Luo and A. K. Sangaiah, "A unified face identification and resolutions scheme using cloud computing in Internet of Things," *Future Generation Computer Systems*, vol. 81, pp. 582–592, 2018.

[47] Z. B. Babovic, V. Protic and V. Milutinovic, "Web performance evaluation for Internet of Things applications," *IEEE Access*, vol. 4, pp. 6974–6992, 2016.

[48] "Internet of Things research study," *Hewlett Packard Enterprise Report*, 2015. http://www8.hp.com/us/en/hp-news/pressrelease.html?id=1909050#.

[49] T. Dierks and C. Allen, "The TLS protocol version 1.0," *IETF RFC*, vol. 2246, 1999. www.ietf.org/rfc/rfc2246.txt.

[50] A. Colacovic and M. Hadzialic, "Internet of Things (IoT): A review of enabling technologies, challenges and open research issues," *Computer Network*, vol. 144, pp. 17–39, 2018.

[51] M. Noura, M. Atiquazzaman and M. Gaedke, "Interoperability in Internet of Things infrastructure: Classification, challenges and future work," in *3rd International Conference, IoTaaS 2017*, Taichung, Taiwan, 20–22 September 2017.

[52] S. G. Tzafestad, "Ethics and law in the Internet of Things world," *Smart Cities*, vol. 1, no. 1, pp. 98–120, 2018.

[53] M. Mosko, I. Solis, E. Uzun and C. Wood, "CCNx 1.0 protocol architecture," *A Xerox Company, Computing Science Laboratory PARC*, Tech. Rep., 2017. http://ccnx.org/pubs/ICNCCNProtocols.pdf.

[54] Y. Wu, J. Li, J. Stankovic, K. Whitehouse, S. Son and K. Kapitanova, "Run time assurance of application-level requirements in wireless sensor networks," in *Proceedings 9th ACM/IEEE International Conference on Information Processing in Sensor Networks*, pp. 197–208, Stockholm, Sweden, 21–16 April 2010.

[55] L. Huo and Z. Wang, "Service composition instantiation based on cross-modified artificial Bee Colony algorithm," *Chin Communications*, vol. 13, no. 10, pp. 233–244, 2016.

[56] G. White, V. Nallur and S. Clarke, "Quality of service approaches in IoT: A systematic mapping," *Journal of Systems and Software*, vol. 132, pp. 186–203, 2017.

[57] ISO/IEC 25010—Systems and software engineering—systems and software quality requirements and evaluation (SQuaRE)—system and software quality models, Technical Report, 2010.

[58] Oasis. Web services quality factors version 1.0.2012. http://docs.oasis-open. org/wsqm/wsqf/v1.0/WS-QualityFactors.pdf

[59] X. Fafoutis et al., "A residential maintenance-free long-term activity monitoring system for healthcare applications," *EURASIP Journal Wireless Communications Network*, 2016. https://doi.org/10.1186/s13638-016-0534-3.

[60] Nate's telesurgery wiki https://sites.google.com/site/telesurgerynk/

[61] C. Vuppalapati, A. Ilapakurti and S. Kedari, "The role of big data in creating sense EHR, an integrated approach to create next generation mobile sensor and wear-able data driven electronic health record (EHR)," in *2016 IEEE 2nd International Conference on Big Data Computing Service and Applications (BigDataService)*, pp. 293–296, New York, 2016.

[62] T. Clausen, U. Herberg and M. Philipp, "A critical evaluation of the IPv6 routing protocol for low power and lossy networks (RPL)," in *2011 IEEE 7th International Conference on Wireless and Mobile Computing, Networking and Communications (WiMob)*, Wuhan, China, 10–12 October 2011.

# Index

Printed in the United States
by Baker & Taylor Publisher Services